SpringerBriefs in Statistics

JSS Research Series in Statistics

The current research of statistics in Japan has expanded in several directions in line with recent trends in academic activities in the area of statistics and statistical sciences over the globe. The core of these research activities in statistics in Japan has been the Japan Statistical Society (JSS). This society, the oldest and largest academic organization for statistics in Japan, was founded in 1931 by a handful of pioneer statisticians and economists and now has a history of about 80 years. Many distinguished scholars have been members, including the influential statistician Hirotugu Akaike, who was a past president of JSS, and the notable mathematician Kiyosi Itô, who was an earlier member of the Institute of Statistical Mathematics (ISM), which has been a closely related organization since the establishment of ISM. The society has two academic journals: the Journal of the Japan Statistical Society (English Series) and the Journal of the Japan Statistical Society (Japanese Series). The membership of JSS consists of researchers, teachers, and professional statisticians in many different fields including mathematics, statistics, engineering, medical sciences, government statistics, economics, business, psychology, education, and many other natural, biological, and social sciences. The JSS Series of Statistics aims to publish recent results of current research activities in the areas of statistics and statistical sciences in Japan that otherwise would not be available in English; they are complementary to the two JSS academic journals, both English and Japanese. Because the scope of a research paper in academic journals inevitably has become narrowly focused and condensed in recent years, this series is intended to fill the gap between academic research activities and the form of a single academic paper. The series will be of great interest to a wide audience of researchers, teachers, professional statisticians, and graduate students in many countries who are interested in statistics and statistical sciences, in statistical theory, and in various areas of statistical applications.

More information about this subseries at http://www.springer.com/series/13497

Hajime Yamato

Statistics Based on Dirichlet Processes and Related Topics

Hajime Yamato (ID)
Kagoshima University
Kagoshima, Japan

ISSN 2191-544X   ISSN 2191-5458   (electronic)
SpringerBriefs in Statistics
ISSN 2364-0057   ISSN 2364-0065   (electronic)
JSS Research Series in Statistics
ISBN 978-981-15-6974-6   ISBN 978-981-15-6975-3   (eBook)
https://doi.org/10.1007/978-981-15-6975-3

This Springer imprint is published by the registered company Springer Nature Singapore Pte Ltd.
The registered company address is: 152 Beach Road, #21-01/04 Gateway East, Singapore 189721,
Singapore

Preface

As a probability distribution on the space of probability distributions, the Dirichlet process was introduced by Ferguson in 1973 and made a new era for nonparametric Bayes inference. Because the Dirichlet process is discrete almost surely, duplications may occur among a sample from it. This property associates the Dirichlet process with the Ewens sampling formula, which was discovered by Ewens relating to genetics in 1972. First, I worked Bayes inference with respect to estimable parameter using the Dirichlet process. This work was very useful for me to study the Ewens sampling formula. This monograph is mainly based on my works on the Dirichlet process and the Ewens sampling formula.

Chapter 1 introduces the Dirichlet process and states its connection with normalized random measure and process neutral to the right. Chapter 2 introduces the relation between the Dirichlet process and the Ewens sampling formula. Its properties are stated using the Chinese restaurant process, along with the Donnelly–Tavare–Griffiths formula I and II. We show the asymptotic distributions of the related statistics. We state also the Pitman sampling formula. Chapter 3 gives a Bayes estimate of estimable parameter using the Dirichlet process as a prior. Its limit is obtained by deleting the effect of a prior. This limit, U-statistic and V-statistic, are written as a convex combination of U-statistics. We show the properties of the convex combination. Chapter 4 gives an approximation to the distribution of number of components for the Ewens sampling formula. With respect to the ordered statistics of a sample from the GEM distribution on the positive integers, its joint and marginal distributions are given. In addition, we introduce the Erdős–Turán law for the Ewens sampling formula. We give a new version of the law for the formula with random parameter.

I would like to thank Professor Masafumi Akahira, this series editor, for giving me the opportunity of writing this monograph and his continuous encouragement. I am also grateful to Professor Taka-aki Shiraisi for his advice on writing the manuscript.

Kagoshima, Japan
2020

Hajime Yamato

Contents

1 **Introduction** 1
   1.1   Dirichlet Process as Prior Distribution in Nonparametric
        Inference 1
   1.2   Normalized Random Measure 2
   1.3   Process Neutral to the Right 3
   References 5

2 **Dirichlet Process, Ewens Sampling Formula, and Chinese**
   Restaurant Process 7
   2.1   Dirichlet Process and Ewens Sampling Formula 7
   2.2   Chinese Restaurant Process 9
   2.3   GEM Distribution 12
   2.4   Poisson–Dirichlet Distribution 13
   2.5   Asymptotic Properties of Statistics Related with ESF 15
        2.5.1   Ewens Sampling Formula 15
        2.5.2   Donnelly–Tavaré–Griffiths Formula I 15
        2.5.3   Donnelly–Tavaré–Griffiths Formula II 16
        2.5.4   Number K_n of Distinct Components (1) 18
        2.5.5   Number K_n of Distinct Components (2) 20
   2.6   Related Topics 21
        2.6.1   Patterns of Communication 21
        2.6.2   Pitman Sampling Formula (1) 22
        2.6.3   Pitman Sampling Formula (2) 23
   References 26

3 **Nonparametric Estimation of Estimable Parameter** 29
   3.1   Nonparametric Bayes Estimator of Estimable Parameter 29
   3.2   Convex Combination of U-Statistics 31
        3.2.1   Preliminaries 31
        3.2.2   Convex Combination of U-Statistics 32

3.3   Properties of Convex Combination of U-Statistics   36
        3.3.1 H-Decomposition.   36
        3.3.2   Berry–Esseen Bound   38
        3.3.3   Asymptotic Distribution for Degenerate Kernel   39
        3.3.4   Corrections to Previous Works   43
3.4   Jackknife Statistic of Convex Combination of U-Statistics   43
References   46

4   **Statistics Related with Ewens Sampling
    Formula**   49
    4.1   Discrete Approximation to Distribution of Number of
            Components   49
            4.1.1   Preliminaries   49
            4.1.2   Ewens Sampling Formula.   54
            4.1.3   Applications   58
    4.2   GEM Distribution on Positive Integers.   59
            4.2.1   Distribution of Statistics from GEM Distribution   59
            4.2.2   Frequency Distribution of Order Statistics from GEM
                     Distribution (1)   65
            4.2.3   Frequency Distribution of Order Statistics from GEM
                     Distribution (2)   67
    4.3   Erdős-Turán Law   68
            4.3.1   Erdős–Turán Law for ESF   68
            4.3.2   Erdős–Turán Law for ESF with Random
                     Parameter   70
    References   71

Index   73

Chapter 1
Introduction

Abstract With respect to Bayesian approach to nonparametric inference, a prior distribution is desirable whose support includes many continuous distributions. Ferguson's Dirichlet process satisfies this property, while a distribution being the Dirichlet process is discrete almost surely. Many works have been done on the Dirichlet process and the related processes. Because the purpose of this monograph is to give some topics associated with the Dirichlet process, we illustrate the process with two related process in this chapter. They are normalized random measure and process neutral to the right.

Keywords Completely random measure · Conjugate · Dirichlet process · Gamma process · Nonparametric inference · Normalized random measure · Prior distribution · Process neutral to the right

1.1 Dirichlet Process as Prior Distribution in Nonparametric Inference

The Dirichlet process was introduced by Ferguson [1] in order to solve nonparametric inference from a Bayesian standpoint. Its definition was given by two forms. One is based on a finite Dirichlet distribution of any finite measurable partition, as a random probability on $(\mathcal{X}, \mathcal{A})$, where $\mathcal{X}$ is a set and $\mathcal{A}$ is a σ-field of subsets of $\mathcal{X}$.

Definition 1.1 ([1]) Let $\alpha(\cdot)$ be a non-null finite measure on $(\mathcal{X}, \mathcal{A})$. P is a Dirichlet process with parameter α and written as $P \in \mathcal{D}(\alpha)$, if for every $m = 1, 2, \ldots$ and measurable partition $\{B_1, \ldots, B_m\}$ of $\mathcal{X}$, the distribution of $(P(B_1), \ldots, P(B_m))$ is the Dirichlet distribution with parameter $(\alpha(B_1), \ldots, \alpha(B_m))$, where measurable partition $(\{B_1, \ldots, B_m\}$ of $\mathcal{X}$ means that each B_i is measurable, disjoint, and $\sum_1^m B_i = \mathcal{X}$.

While this definition of the Dirichlet process is based on a finite distribution of any finite partition, the process gives a random probability measure on $(\mathcal{X}, \mathcal{A})$ because the Kolmogorov consistency condition is satisfied.

© The Author(s), under exclusive license to Springer Nature Singapore Pte Ltd. 2020
H. Yamato, *Statistics Based on Dirichlet Processes and Related Topics*,
JSS Research Series in Statistics, https://doi.org/10.1007/978-981-15-6975-3_1

In this monograph, we consider the Dirichlet process and its related processes on $(\mathbb{R}, \mathcal{B})$, where $\mathbb{R}$ is the real line and $\mathcal{B}$ is its Borel σ-field.

The other definition of the Dirichlet process is related with a gamma process which is based on the previous work Ferguson and Klass [2]. The equivalent definition is given in Proposition 2.3, which is quoted from (78) of Kingman [3].

A distribution being the Dirichlet process is discrete almost surely (a.s.). Maybe it seems that the Dirichlet process is not suitable for a prior distribution of nonparametric inference. But the support of the Dirichlet process with parameter α contains the set of all probability distributions absolutely continuous with respect to α, as stated in the first paragraph of p. 216 of [1]. For example, if $\alpha(\cdot)/\alpha(\mathbb{R})$ is Student's t-distribution, then the support of $\mathcal{D}(\alpha)$ contains every continuous distribution.

As a prior distribution, its conjugacy property is appropriate. The Dirichlet process is conjugate, that is, the conditional distribution given a sample from Dirichlet process is also the Dirichlet process.

We show two classes of random probability measures related to the Dirichlet process, which are normalized random measure and process neutral to the right.

1.2 Normalized Random Measure

Completely random measure was introduced by Kingman [4] in its p. 60–61.

Definition 1.2 ([4]) A random measure $\tilde{\mu}$ on $(\mathbb{R}, \mathcal{B})$ is said to be completely random measure, if (i) $\tilde{\mu}(\emptyset) = 0$ almost surely (a.s.) and (ii) for any finite disjoint sets $A_1, A_2, \cdots, A_n (\in \mathcal{B})$, the random variables $\tilde{\mu}(A_1), \tilde{\mu}(A_2), \ldots \tilde{\mu}(A_m)$ are independent.

Any completely random measure consists of deterministic measures and a purely atomic random measure. We consider the completely random measure only consisting of the later purely atomic random measure. Then, for any $B (\in \mathcal{B})$, the Laplace transform of $\tilde{\mu}(B)$ is

$$E\left[e^{-\lambda\tilde{\mu}(B)}\right] = \exp\left\{-\int_{\mathbb{R}^+ \times B} [1 - e^{-\lambda v}]\nu(dv, dx)\right\}, \quad \lambda > 0, \qquad (1.1)$$

where $\mathbb{R}^+$ is the positive real line. Normalized random measure with independent increments (NRMI) was introduced by Regazzini et al. [5, (2)]. The following is quoted from Definition 2 of Lijoi et al. [6].

Definition 1.3 ([5, 6]) Assume that $\tilde{\mu}$ is completely random measure with intensity measure ν. Then,

$$\tilde{P}(\cdot) = \frac{\tilde{\mu}(\cdot)}{\tilde{\mu}(\mathbb{R})} \qquad (1.2)$$

is said to be NRMI on $(\mathbb{R}, \mathcal{B})$, where $\psi_{\mathbb{R}}(\lambda) < \infty$ for any $\lambda > 0$ and $\nu(\mathbb{R}^+, \mathbb{R}) = \infty$.

The assumptions $\psi_{\mathbb{R}}(\lambda) < \infty$ and $\nu(\mathbb{R}^+ \times \mathbb{R}) = +\infty$ ensure $0 < \tilde{\mu}(\mathbb{R}) < \infty$ a.s. $\tilde{\mu}$ and NRMI $\tilde{P}$ is said to be homogeneous, if $\nu(dv, dx) = \rho(dv)\alpha(dx)$ with measures ρ on $\mathbb{R}^+$ and α on $\mathbb{R}$. For example, NRMI with the intensity measure,

$$\nu(dv, dx) = \frac{e^{-\tau v}}{\Gamma(1 - \sigma)v^{1+\sigma}} dv\alpha(dx) ,$$

is said to be a generalized gamma NRMI, where $\sigma < 1$ and $\tau \geq 0$ (Brix [7, (2)]). Taking $\sigma = 0$ and $\tau = 1$, we have the intensity measure

$$\nu(dv, dx) = \frac{e^{-v}}{v} dv\alpha(dx)$$

of the gamma process. NRMI associated with this intensity measure is the Dirichlet process with parameter α (Kingman [8, 9.2]). In terms of conjugacy, the following characterization of the Dirichlet process was given by Theorem 1 of James et al. [9].

Proposition 1.1 ([9]) *Let $\mathcal{P}$ be the class of homogeneous NRMI and let $\tilde{P}$ be in $\mathcal{P}$. Then, the posterior distribution of $\tilde{P}$, given a sample $X_1, \ldots, X_n$, is in $\mathcal{P}$ if and only if $\tilde{P}$ is the Dirichlet process.*

1.3 Process Neutral to the Right

Process neutral to the right was introduced by Doksum [10, Definition 3.1].

Definition 1.4 ([10]) A random distribution function F on $\mathbb{R}$ is said to be neutral to the right if for every positive integer m and $t_1 < t_2 < \cdots < t_m$, there exist independent random variables $V_1, V_2, \ldots, V_m$ such that

$$\left(1 - F(t_1), 1 - F(t_2), \ldots, 1 - F(t_m)\right) \stackrel{d}{=} (V_1, V_1 V_2, \ldots, V_1 V_2 \cdots V_m) . \quad (1.3)$$

The condition for neutral to the right is essentially equivalent to that of the random variables

$$1 - F(t_1), \frac{F(t_2) - F(t_1)}{1 - F(t_1)}, \ldots, \frac{F(t_m) - F(t_{m-1})}{1 - F(t_{m-1})}$$

or

$$1 - F(t_1), \frac{1 - F(t_2)}{1 - F(t_1)}, \ldots, \frac{1 - F(t_m)}{1 - F(t_{m-1})}$$

which are independent, except for the possibility that the denominators are equal to zero ([10, (1.1)] and Ferguson [11, p. 623, ↓ 4–5]).

If P is the Dirichlet process with parameter α on $(\mathbb{R}, \mathcal{B})$, then the distribution of $\left(P(-\infty, t_1), P(t_1, t_2), \ldots, P(t_{m-1}, t_m), P(t_m, \infty)\right)$ is the Dirichlet distribution with parameter $\big(\alpha(-\infty, t_1), \alpha(t_1, t_2), \ldots, \alpha(t_{m-1}, t_m), \alpha(t_m, \infty)\big)$, where $t_1 < t_2 < \cdots < t_m$. Thus, the relation (1.3) holds with the independent random variables $V_1, V_2, \ldots, V_m$ such that V_i has beta $(\alpha(t_{i-1}, t_i], \alpha(t_i, \infty))$ for $(i = 1, 2, \ldots, m)$. Therefore, the Dirichlet process is neutral to the right.

The class of processes neutral to the right process includes the Dirichlet process. The process neutral to the right has a simple expression.

Proposition 1.2 ([10]) *A random distribution function F on $\mathbb{R}$ neutral to the right if and only if it has the same probability distribution as*

$$1 - \exp[-Y_t]$$

for some non-decreasing, right continuous a.s., independent process Y_t with $\lim_{t \to -\infty} Y_t = 0$ and $\lim_{t \to \infty} Y_t = \infty$ a.s.

In other words, Y_t is equal to $\tilde{\mu}(-\infty, t]$ for some completely random measure $\tilde{\mu}$ on $(\mathbb{R}, \mathcal{B})$. Therefore, as a corollary to Proposition 1.2, the discreteness of the process is obtained.

Corollary 1.1 ([10, Corollary 3.2]) *If the random distribution function F neutral to the right and if $-\log\big(1 - F(t)\big)$ has no nonrandom part, then*

$$F \text{ is a discrete distribution function a.s.}$$

The process has the conjugacy.

Proposition 1.3 ([10, Theorem 4.2]) *If $X_1, \ldots, X_n$ is a sample from process neutral to the right, then the posterior distribution of P given $X_1, \ldots, X_n$ is neutral to the right.*

Ferguson and Phadia [12] gave the following result for right-censored date in Theorem 3.

Proposition 1.4 ([12]) *Let F be a random distribution function neutral to the right. Let X be a sample of size one from F and let x be a number in $(0, \infty)$. Then,*

1. *the posterior distribution of F given $X > x$ is neutral to the right and*
2. *the posterior distribution of F given $X \geq x$ is neutral to the right.*

While we state that the Dirichlet process is a process neutral to the right in the second paragraph, we see this relation in terms of completely random measure on $\mathbb{R}^+$. For a process $F(t)$ neutral to the right with $F(0) = 0$, $Y_t = -\log(1 - F(t))$ is an increasing Levy process such that $Y_0 = 0$ and $\lim_{t \to \infty} Y_t = \infty$. Put $\tilde{\mu}(0, t] := Y_t$. Then, $\tilde{\mu}$ is a completely random measure on $\mathbb{R}^+$. Following James [13, the last paragraph of p. 39], Lijoi and Prünster [14, Example 3.10]), we consider $\tilde{\mu}$ having intensity measure

$$\nu(dv, dx) = \frac{e^{-v\alpha(x, \infty)}}{1 - e^{-v}} \alpha(dx) dv ,$$

where α is a measure on $\mathbb{R}^+$. Then, the Laplace transform of $\tilde{\mu}(0, t])(= Y_t)$ is

$$E\left[e^{-\lambda\tilde{\mu}((0,t])}\right] = \exp\left\{-\int\limits_0^\infty \int\limits_0^t [1 - e^{-\lambda v}]\frac{e^{-v\alpha(x,\infty)}}{1 - e^{-v}}\alpha(dx)dv\right\}$$

$$= \exp\left\{-\int\limits_0^\infty \frac{e^{-v\alpha(\mathbb{R}^+)}(e^{v\alpha(0,t]} - 1)}{(1 - e^{-v})v}dv\right\},$$

which corresponds with the Laplace transform of $Y_t' = -\log(1 - F'(t))$ with $F' \in \mathcal{D}(\alpha)$ ([11, p. 626, ↓ 1–4]). Therefore, $F(t) = 1 - e^{-\tilde{\mu}(0,t]}(= 1 - e^{-Y_t})$ is the Dirichlet process with parameter α on $\mathbb{R}^+$.

References

1. Ferguson, T.S.: A Bayesian analysis of some nonparametric problems. Ann. Stat. **1**, 209–230 (1973)
2. Ferguson, T.S., Klass, M.J.: A representation of independent increment processes without Gaussian components. Ann. Math. Stat. **43**, 1634–1643 (1972)
3. Kingman, J.F.C.: Random discrete distributions. J. R. Stat. Soc. B, **37**, 1–15 (1975)
4. Kingman, J.F.C.: Completely random measures. Pac. J. Math. **21**, 59–78 (1969)
5. Regazzini, E., Lijoi, A., Prünster, I.: Distributional results for means of normalized random measures with independent increments. Ann. Stat. **31**, 560–585 (2003)
6. Lijoi, A., Prünster, I., Walker, S.G.: Investigating nonparametric priors with Gibbs structure. Stat. Sinica. **18**, 1653–1668 (2008)
7. Brix, A.: Generalized gamma measures and shot-noise Cox processes. Adv. Appl. Probab. **31**, 929–953 (1999)
8. Kingman, J.F.C.: Poisson Processes. Oxford University Press, New York (1993)
9. James, L., Lijoi, A., Prünster, I.: Conjugacy as a distinctive feature of the Dirichlet process. Scandinavian J. Stat. **33**, 105–120 (2006)
10. Doksum, K.: Tailfree and neutral random probabilities and their posterior distributions. Ann. Probab. **2**, 183–201 (1974)
11. Ferguson, T.S.: Prior distribution on spaces of probability measures. Ann. Stat. **2**, 615–629 (1974)
12. Ferguson, T.S., Phadia, E.G.: Bayesian nonparametric estimation based on censored data. Ann. Stat. **7**, 163–186 (1979)
13. James, L.F.: Poisson process partition calculus with application to exchangeable models and Bayesian nonparametrics (2002). arXiv: math/0205093v1
14. Lijoi, A., Prünster, I.: Model beyond the Dirichlet process. In: Hjort. N.L., Holmes, C., Müller, P., Walker, S. G. (eds.) Bayesian Nonparametrics, pp. 80–136. Cambridge University Press, Cambridge (2010)

Chapter 2
Dirichlet Process, Ewens Sampling Formula, and Chinese Restaurant Process

Abstract The Dirichlet process is a random probability measure and its realization is discrete almost surely. Therefore, there may be duplications among a sample from a distribution having the Dirichlet process. The distribution of this duplication is well known as the Ewens sampling formula. This formula is derived by another model, for example, the Chinese restaurant process. The Ewens sampling formula is related with the Donnelly–Tavaré–Griffiths formula I and II, the GEM distribution, and the Poisson–Dirichlet distribution. The Donnelly–Tavaré–Griffiths formula II is related with the Yule distribution and the Waring distribution. The distribution of the number of distinct components of the Ewens sampling formula asymptotically converges to normal distribution. It converges also to the shifted Poisson distribution under the condition resembled to that of Poisson law of small number. As a formula related to the Ewens sampling formula, the Pitman sampling formula is well known. It is also derived by the Chinese restaurant process.

Keywords Chinese restaurant process · Dirichlet process · Donnelly–Tavaré–Griffiths formula · Ewens sampling Formula · GEM distribution · Pitman sampling formula · Poisson–Dirichlet distribution · Yule distribution · Waring distribution

2.1 Dirichlet Process and Ewens Sampling Formula

As stated in Chap. 1, we consider the Dirichlet process on $(\mathbb{R}, \mathcal{B})$. We put $\theta = \alpha(\mathbb{R})$ and $Q(\cdot) = \alpha(\cdot)/\alpha(\mathbb{R})$. In order to denote that P is Dirichlet process on $(\mathbb{R}, \mathcal{B})$ with parameters θ and Q, we use $P \in \mathcal{D}(\theta, Q)$, instead of $P \in \mathcal{D}(\alpha)$.

For Dirichlet process introduced by Ferguson [1], its alternative form is given by Sethuraman [2]. Let the sequence of random variables $V_j (j = 1, 2, \ldots)$ be independent and identically distributed with the distribution Q. Let the sequence of random variables $W_j (j = 1, 2, \ldots)$ be independent and identically distributed with the beta distribution beta $(1, \theta)$ whose density is $\theta(1-x)^{\theta-1}$ $(0 < x < 1; \theta > 0)$. Suppose that $V_j (j = 1, 2, \ldots)$ and $W_j (j = 1, 2, \ldots)$ are independent. Put

© The Author(s), under exclusive license to Springer Nature Singapore Pte Ltd. 2020
H. Yamato, *Statistics Based on Dirichlet Processes and Related Topics*,
JSS Research Series in Statistics, https://doi.org/10.1007/978-981-15-6975-3_2

$$p_1 = W_1, \quad p_j = W_j(1 - W_1) \cdots (1 - W_{j-1}) \ (j = 2, 3, \ldots) . \qquad (2.1)$$

Then, P having the Dirichlet process $\mathcal{D}(\theta, Q)$ can be written as follows:

$$P(A) = \sum_{j=1}^{\infty} p_j \delta_{v_j}(A) \quad (A \in \mathcal{B}) . \qquad (2.2)$$

The distribution of the random variables

$$\mathbf{P}_G = (p_1, p_2, \ldots)$$

is said the GEM distribution GEM (θ) with parameter θ. This is named after Griffiths (unpublished note), Engen [3], and MacCloskey [4] by Ewens [5, p. 217]. From (2.2), we see the well-known fact that P is discrete a.s., even if Q is continuous.

For the sample $X_1, \ldots, X_n$ from $P \in \mathcal{D}(\theta, Q)$ with a continuous Q, let C_1 denote the number of distinct values that occur only once, C_2 that occur exactly twice, ..., C_n that occur exactly n times. By (1.5), for $c_1, c_2, \ldots, c_n$ such that $\sum_{i=1}^{n} i c_i = n$ and $\sum_{i=1}^{n} c_i = k$,

$$P\big((C_1, C_2, \ldots, C_n) = (c_1, c_2, \ldots, c_n)\big) = \frac{\theta^k}{\theta^{[n]}} \cdot \frac{n!}{\prod_{j=1}^{k} j^{c_j} c_j!} \quad (0 < \theta < \infty) , \qquad (2.3)$$

where $\theta^{[n]} = \theta(\theta + 1) \ldots (\theta + n - 1)$. This formula was derived by Antoniak [6] in the context of Bayesian statistics. It was also independently derived by Ewens [7] in the context of genetics and is well known as the Ewens sampling formula (ESF). The formula appears in many statistical contexts, including permutations, ecology, and patterns of communication (e.g., see Johnson et al. [8, Chap. 41]). Let K_n denote the number of distinct observations. Then we have $K_n = \sum_{i=1}^{n} C_i$. The probability function (p.f) of K_n is

$$(K_n = k) = \mid s(n, k) \mid \frac{\theta^k}{\theta^{[n]}} \quad (k = 1, 2, \ldots, n) , \qquad (2.4)$$

where $\mid s(n, k) \mid$ is the signless Stirling number of the first kind [7].

In addition, the Dirichlet process has the following property.

Theorem 2.1 ([1, Theorem 1]) *If $P \in \mathcal{D}(\theta, Q)$ and if $X_1, \ldots, X_n$ is a sample from P, then the posterior distribution of P given $X_1, \ldots, X_n$ is $P \in \mathcal{D}\big(\theta + n, (\theta Q + \sum_1^n \delta_{x_i})/(\theta + n)\big)$, where δ_x is the measure giving mass one to x.*

Thus we have

$$P(X_{n+1} \in A \mid X_1, \ldots, X_n) = \sum_{j=1}^{n} \frac{1}{\theta + n} \delta_{X_j}(A) + \frac{\theta}{\theta + n} Q(A), \text{ for } A \in \mathcal{B} . \qquad (2.5)$$

This is a Pólya sequence with parameter (θ, Q) (Blackwell and MacQueen [9]). Under the assumption that the distribution Q is continuous, X_n takes a same value as the previous one with the probability $1/(\theta + n)$ or a new value with the probability $\theta/(\theta + n)$. If we consider only duplication among the sample $X_1, \ldots, X_n$, this sequential relation (2.5) is equivalently described by the Chinese restaurant process (e.g., see Aldous [10, 11.19] and Pitman [11, p. 59]). The equivalent model is also given by the pattern of communication (Taga and Isii [12]) which is explained in Sect. 2.6.1. Related to the relation (2.5), Lo [13] gave the following property of the Dirichlet process.

Proposition 2.1 ([13]) *Let P be a random probability on $(\mathbb{R}, \mathcal{B})$. If for each $n = 1, 2, \ldots$, there is a constant $a_n \in (0, 1)$ and a probability P_n on $(\mathbb{R}, \mathcal{B})$ such that the posterior mean of P given a sample $X_1, \ldots, X_n$ from P is*

$$\frac{a_n}{n} \sum_{i=1}^{n} \delta_{X_i} + (1 - a_n) P_n , \tag{2.6}$$

then

1. *$P_n = P_0$ for $n = 1.2, \ldots$, where P_0 is $E(P)$, defined by $P_0(B) = E[P(B)]$ for each $B \in \mathcal{B}$,*
2. *$a_n = n / (a_1^{-1} - 1 + n)$ for $n = 1, 2, \ldots$, and*
3. *P is a Dirichlet process $\mathcal{D}(a_1^{-1} - 1, P_0)$.*

Then, by this proposition and Theorem 2.1, the following was obtained.

Corollary 2.1 ([13]) *P is a Dirichlet process on $(\mathbb{R}, \mathcal{B})$ if and only if for each $n = 1, 2, \ldots$, the posterior mean of P given a sample $X_1, \ldots, X_n$ from P is given by (2.6), for some $a_n \in (0, 1)$ and some probability P_n on $(\mathbb{R}, \mathcal{B})$.*

2.2 Chinese Restaurant Process

We explain $(C_1, C_2, \ldots, C_n)$ and K_n using the Chinese restaurant process for two forms (I) and (II):
The Chinese restaurant process (I): Consider customers $1, 2, \ldots, n$ arriving sequentially at an initially empty restaurant that has many tables.

(i) The first customer sits at the first table.
(ii) The j th customer selects a table as follows ($j = 2, 3, \ldots, n$): the customer selects

$$\text{(1) a new table with probability } \frac{\theta}{\theta + j - 1},$$

$$\text{(2) an occupied table with probability } \frac{j - 1}{\theta + j - 1}.$$

Define random variables ξ_j $(j = 1, \ldots, n)$ as follows:

$\xi_j := 1$ if the jth customer sits at a new table,

$\xi_j := 0$ if the jth customer sits at an occupied table.

Then, we have

$$P(\xi_j = 1) = \frac{\theta}{\theta + j - 1}, \quad P(\xi_j = 0) = \frac{j - 1}{\theta + j - 1} \quad (j = 1, \ldots, n) . \quad (2.7)$$

After n customers are seated, the number of occupied tables is given by

$$K_n = \xi_1 + \xi_2 + \cdots + \xi_n , \quad (2.8)$$

with a p.f. given by (2.4).

Here, we consider the ordered tables. Let $A_1^*, A_2^*, \ldots$ denote the set of customers occupying the first table, second table, and so on. For each table, only its first customer is specified. For a disjoint partition $a_1^*, a_2^*, \ldots, a_k^*$ such that $a_1^* \cup a_2^* \cup \cdots \cup a_k^* = \{1, 2, \ldots, n\}$, we have

$$P((A_1^*, A_2^*, \ldots, A_k^*) = (a_1^*, a_2^*, \ldots, a_k^*)) = \frac{\theta^k}{\theta^{[n]}} \prod_{j=1}^k (| a_j^* | - 1)! , \quad (2.9)$$

where $| a^* |$ is the cardinal of set a^*. Let $A_1, A_2, \ldots$ denote the number of customers occupying the first table, second table, and so on. Multiplying the binomial coefficient $\binom{n-1}{a_1-1}\binom{n-a_1-1}{a_2-1} \cdots \binom{n-a_1-\cdots-a_{k-1}-1}{a_k-1}$ by (2.9), we have

$$P(A_1 = a_1, A_2 = a_2, \ldots, A_k = a_k)$$
$$= \frac{\theta^k}{\theta^{[n]}} \cdot \frac{n!}{a_k(a_k + a_{k-1}) \cdots (a_k + a_{k-1} + \cdots + a_1)} , \quad (2.10)$$

for the positive integers $a_1, a_2, \ldots, a_k$ satisfying $a_1 + a_2 + \cdots + a_k = n$. Tavaré [14] and Joyce and Tavaré [15] derived this distribution using processes with immigration. Let

C_1 be the number of tables occupied by one customer,

C_2 be the number of tables occupied by two customers, and so on.

Equation (2.10) relates with the ESF. Using (2.15) of Remark 2.1 to (2.10), we know that these $(C_1, \ldots, C_n)$, corresponding to $(A_1, \ldots, A_k)$, have ESF(θ).

The Chinese restaurant process (II): The $0-1$ sequence $\xi_1 \xi_2 \ldots \xi_n 1$ $(\xi_1 = 1)$ indicates that each customer $(1, 2, \ldots, n)$ sits at a new table (1) or at an occupied table (0). Let

$$C_j := \text{no. of } j - \text{spacings in } \xi_1 \xi_2 \ldots \xi_n 1 \quad (j = 1, \ldots, n) ,$$

which can be written as

$$C_1 = \sum_{i=1}^{n-1} \xi_i \xi_{i+1} + \xi_n , \quad C_n = \xi_1 \bar{\xi}_2 \cdots \bar{\xi}_n , \tag{2.11}$$

$$C_j = \sum_{i=1}^{n-j} \xi_i \bar{\xi}_{i+1} \cdots \bar{\xi}_{i+j-1} \xi_{i+j} + \xi_{n-j+1} \bar{\xi}_{n-j+2} \cdots \bar{\xi}_n \quad (j = 2, \ldots, n-1) , \tag{2.12}$$

where $\bar{\xi}_i = 1 - \xi_i$ $(i = 1, 2, \ldots, n)$. Then,

$$K_n = C_1 + C_2 + \cdots + C_n , \tag{2.13}$$

(see, Arratia et al. [16, p. 523]).

Here, we consider the ordered spacings. Let $D_1, D_2, \ldots$ be the length of the first spacing, second spacing, and so on. The ordered spacing $(D_1, D_2, \ldots, D_k)$ is based on the $0-1$ sequence $\xi_1 \xi_2 \ldots \xi_n 1$ $(\xi_1 = 1)$. Its reversed sequence $1\xi_n \xi_{n-1} \ldots \xi_1$ gives the reversed spacing $(D_k, D_{k-1}, \ldots, D_1)$, and is associated with the Feller coupling (e.g., see [16, p. 523]). Using the property of the Feller coupling to the left-hand side of

$$P(D_1 = d_1, D_2 = d_2, \ldots, D_k = d_k) = P(D_k = d_k, D_{k-1} = d_{k-1}, \ldots, D_1 = d_1) ,$$

we have

$$P(D_1 = d_1, D_2 = d_2, \ldots, D_k = d_k) = \frac{\theta^k}{\theta^{[n]}} \cdot \frac{n!}{d_1(d_1 + d_2) \cdots (d_1 + d_2 + \cdots + d_k)} , \tag{2.14}$$

for the positive integers $d_1, d_2, \ldots, d_k$ satisfying $d_1 + d_2 + \cdots + d_k = n$ (Yamato [17, Appendix]). Then, by (2.15), $(C_1, C_2, \ldots, C_n)$, corresponding to $(D_1, \ldots, D_k)$, has ESF(θ). This correspondence is also stated in Donnelly and Tavaré [18, p. 10], where their $D_1, D_2, \ldots$ denote waiting times of the next mutation in their coalescent process. For the property that $(C_1, C_2, \ldots, C_n)$, given by (2.11) and (2.12), has ESF(θ), see for example, Arratia and Tavaré [19, Theorem 1]) and Arratia et al. [20, 2nd paragraph]). The distribution (2.13) is also given by Ethier [21, Theorem 4.1].

Ewens [5, p. 221] refers to the distribution with form (2.10) as the Donnelly–Tavaré–Griffiths formula, quoting his relation (163), which is equivalent to (2.10). However, the distribution with form (2.14) is stated in the cited reference, [18, (4.2)]). Therefore, Yamato [22] refers to the distributions with forms (2.10) and (2.14) as the Donnelly–Tavaré–Griffiths formula I and II, respectively.

Remark 2.1 Let $(C_1, C_2, \ldots, C_n)$ be a random partition of a positive integer n satisfying $\sum_{i=1}^{n} iC_i = n$. We first consider the condition that $(C_1, C_2, \ldots, C_n)$ is exchangeable. That is, the distribution of $(C_1, C_2, \ldots, C_n)$ is invariant under any permutation of $\{1, 2, \ldots, n\}$. Nacu [23, Theorem 2] showed the following:

Proposition 2.2 ([23]) *The Ewens sampling formula associated with the Chinese restaurant process* (II) *is the only exchangeable partition with independent 0–1 sequence.*

Young [24, Theorem 2] showed this property, by adding partition structure to the exchangeability. The partition structure is due to Kingman [25]. It means consistency of the distribution of P_n of $(C_1, C_2, \ldots, C_n)$. That is, if an object is deleted uniformly at random from $(C_1, C_2, \ldots, C_n)$, then P_{n-1} is the same type as P_n (Pitman [26]). Young [24] showed that convolution of 0–1 sequences of the Chinese restaurant process (II) and the discrete homogeneous renewal sequence $p_n = (-1)^{n-1}\binom{\alpha}{n}$ ($n \geq 1$, $0 < \alpha < 1$) yields the Pitman sampling formula (2,42).

Remark 2.2 As a special case of [18, Proposition 2.1], we consider the case in which duplications exist among $\mu_1, \mu_2, \ldots, \mu_l$, and let $b_1, \ldots, b_k$ be their distinct values. Let c_j be the number of μ_i equal to b_j, for $j = 1, \ldots, k$ and $i = 1, \ldots, l$. Let Π^* be the collection of permutations $\mathbf{\beta} = \big(\pi(1), \ldots, \pi(l)\big)$, excluding the duplications of $\mu_1, \mu_2, \ldots, \mu_l$.

Then, we have the following:

Lemma 2.1

$$\sum_{\mathbf{\beta} \in \Pi^*} \frac{1}{\mu_{\pi(1)}(\mu_{\pi(1)} + \mu_{\pi(2)}) \cdots (\mu_{\pi(1)} + \mu_{\pi(2)} + \cdots + \mu_{\pi(l)})}$$

$$= \frac{1}{\mu_1 \cdots \mu_l c_1! \cdots c_k!} = \prod_{j=1}^{k} \frac{1}{b_j{}^{c_j} c_j!}. \qquad (2.15)$$

2.3 GEM Distribution

Let $p_1, p_2, \ldots$ be a sequence of random variables such that $0 < p_j < 1$ ($j = 1, 2, \ldots$) and $\sum_{j=1}^{\infty} p_j = 1$. We put its residual fractions as follows:

$$W_1 := p_1, \quad W_j := \frac{p_j}{1 - p_1 - \cdots - p_{j-1}} \quad (j = 3, 4, \ldots). \qquad (2.16)$$

Then, we have the following expression as same as (1.1):

$$p_1 = W_1, \quad p_j = W_j(1 - W_1) \cdots (1 - W_{j-1}) \ (j = 2, 3, \ldots).$$

Then, $\mathbf{P} = (p_1, p_2, \ldots)$ with this form is well known as residual allocation model, stick-breaking model (Halmos [27], Patil and Taillie [28]) or Bernoulli sieve (Gnedin [29]).

We consider the random rearrangement of $\mathbf{P} = (p_1, p_2, \ldots)$, which is obtained from size-biased random permutation. Let $(\tilde{p}_1, \tilde{p}_2, \tilde{p}_3, \ldots)$ be the size-biased permutation of $(p_1, p_2, p_3, \ldots)$. That is, its probabilities are given by

$$P(\tilde{p}_1 = p_{i_1} \mid p_1, p_2, p_3, \ldots) = p_{i_1},$$

and for $j = 2, 3, \ldots$

$$P(\tilde{p}_1 = p_{i_1}, \tilde{p}_2 = p_{i_2}, \ldots, \tilde{p}_j = p_{i_j} \mid p_1, p_2, p_3, \ldots)$$
$$= p_{i_1} \cdot \frac{p_{i_2}}{1 - p_{i_1}} \cdots \frac{p_{i_j}}{1 - p_{i_1} - p_{i_2} - \cdots - p_{i_{j-1}}}.$$

For size-biased permutation, the following property is well known (i.e., see [8] and Pitman [30, Theorem 2]).

Lemma 2.2 *For $(p_1, p_2, p_3, \ldots)$ having the GEM distribution, its size-biased permutation $(\tilde{p}_1, \tilde{p}_2, \tilde{p}_3, \ldots)$ has the same GEM distribution.*

Using this invariance of the GEM distribution under size-biased permutation, the following lemma is obtained.

Lemma 2.3 *Suppose that $\mathbf{P}_G = (p_1, p_2, \ldots)$ has GEM(θ). For the partition $(r_1, \ldots, r_j)$ $(r_1 + \cdots + r_j = k, \ r_1, \ldots, r_j > 0)$ of k,*

$$E\left\{ \sum_{i_1 \neq i_2 \neq \cdots \neq i_j} p_{i_1}^{r_1} p_{i_2}^{r_2} \cdots p_{i_j}^{r_j} \right\} = \frac{\theta^j}{\theta^{[k]}} \prod_{i=1}^{j} (r_i - 1)!. \tag{2.17}$$

(see, Sibuya and Yamato [31, Lemmas 3 and 4] and Yamato [32, A.3]. The right-hand side of (2.17) is equivalent to the probability (2.9). In addition, for the sample $X_1, X_2, \ldots, X_n$ from $P \in \mathcal{D}(\theta, Q)$ with a continuous Q, the probability

$$P(X_1 = \cdots = X_{r_1} \neq X_{r_1+1} = \cdots = X_{r_1+r_2} \neq \cdots, \neq X_{n-r_k+1} = \cdots = X_n)$$

is equal to the right-hand side of (2.17).

2.4 Poisson–Dirichlet Distribution

For $(p_1, p_2, \ldots)$ having the GEM distribution GEM(θ), its rearrangement $(\hat{p}_1, \hat{p}_2, \ldots)$ by decreasing order has the Poisson–Dirichlet distribution PD(θ) which was introduced by Kingman [33].

Let $\{\gamma_t, t \geq 0\}$ be the standard gamma process. γ_t has the standard gamma distribution whose density is given by

$$f(x) = \frac{1}{\Gamma(t)} x^{t-1} e^{-x} \quad (x > 0).\tag{2.18}$$

For the Laplace transformation $E[e^{-u\gamma_t}] = e^{-t\psi(u)}$ of γ_t, it holds that

$$\psi(u) = \int_0^\infty (1 - e^{-ux}) \Lambda(dx), \quad \Lambda(dx) = x^{-1} e^{-x} dx .$$

The jump $\{\Delta_s = \gamma_s - \gamma_{s-}, s > 0\}$ is the Poisson process with intensity measure $\Lambda(dx)$ and for $t > 0$,

$$\gamma_t = \sum_{o < s \le t} \Delta_s \quad (\Delta_s = \gamma_s - \gamma_{s-}).$$

The process increases only at discrete points. For $\theta > 0$, by ordering the jump Δ_s over the interval $[0, \theta]$ of the gamma process $\{\gamma_t, t \ge 0\}$, we put

$$J_1(\gamma_\theta) \ge J_2(\gamma_\theta) \ge J_3(\gamma_\theta) \ge \cdots \ge 0,$$

and

$$\hat{p}_i = \frac{J_i(\gamma_\theta)}{\gamma_\theta} \quad (i = 1, 2, \ldots).$$

Then,

Proposition 2.3 ([33, 34])

$(\hat{p}_1, \hat{p}_2, \ldots)$ *has the Poisson–Dirichlet distribution* $PD(\theta)$.

In addition, Kingman [33] gave an alternative representation of the Dirichlet process: Suppose that a random probability $\mathbf{m}$ such that (i) the values of $\mathbf{m}$ on disjoint sets are independent and (ii) the density of $\mathbf{m}(A)$ ($A \in \mathcal{B}$) is given by (2.18) with $t = \theta Q(A)$. Then,

Proposition 2.4 ([33, 34])

$$P(A) = \frac{\mathbf{m}(A)}{\mathbf{m}(\mathbb{R})} \quad (A \in \mathcal{B})$$

is the Dirichlet process $\mathcal{D}(\theta, Q)$.

Kingman [34, p. 98, ↓ 12–13] said that the Poisson–Dirichlet distribution is rather less than user-friendly. In addition, he said that another distribution of simpler form is the GEM distribution, which is obtained from the Poisson–Dirichlet distribution by size-biased sampling.

2.5 Asymptotic Properties of Statistics Related with ESF

2.5.1 Ewens Sampling Formula

Let Z_i $(i = 1, 2, \ldots)$ be independent random variables with the Poisson distribution of parameter θ/i, $Po(\theta/i)$. For $(C_1, C_2, \ldots, C_n)$ having ESF(θ), its distribution is expressed by the conditional probability using Z_i $(i = 1, 2, \ldots, n)$ as follows:

$$\mathcal{L}(C_1, C_2, \ldots, C_n) = \mathcal{L}\big((Z_1, Z_2, \ldots, Z_n) \mid Z_1 + 2Z_2 + \cdots + nZ_n = n\big),$$

where $\mathcal{L}(C)$ denotes the distribution of C. Using this relation, the asymptotic distribution of $(C_1, C_2, \ldots, C_n)$ was derived by [16, Theorem 1].

Proposition 2.5 ([16])

$$(C_1, C_2, \ldots) \xrightarrow{d} (Z_1, Z_2, \ldots) \quad \text{as } n \to \infty,$$

where $\xrightarrow{d}$ denotes convergence in distribution. This result can be also derived using moments which were given by Watterson [35].

2.5.2 Donnelly–Tavaré–Griffiths Formula I

For $(A_1, A_2, \ldots)$ having the Donnelly–Tavaré–Griffiths formula I given by (2.10), we have

Lemma 2.4 ([31, Lemma 1]) *For a positive a,*

$$P(A_1 = a) = \theta \cdot \theta^{[n-a]}(n-1)^{(c-1)}/\theta^{[n]},$$

where $x^{(r)} = x(x-1)\ldots(x-r+1)$. For a positive integer $m(< n)$ and positive integers $a_1, \ldots, a_m$ $(a_1 + \cdots + a_m < n)$

$$P(A_1 = a_1, \ldots, A_m = a_m) = \frac{\theta^m \cdot \theta^{[n-a_1-\cdots-a_m]}(n-1)^{(a_1+\cdots+a_m-1)}}{\theta^{[n]}(n-a_1)\cdots(n-a_1-\cdots-a_{m-1})}.$$

That is, $A_n - 1$ has the negative hypergeometric distribution (or the beta binomial distribution) NgHg $(n-1, 1, \theta)$ such that

$$P(A_1 - 1 = x) = \binom{-1}{x}\binom{-\theta}{n-1-x}\bigg/\binom{-\theta-1}{n-1}.$$

Under the condition that $A_1 = a_1, \ldots, A_{m-1} = a_{m-1}$ are given, A_m has NgHg $(n - a_1 dots - a_{m-1}, 1, \theta)$.

This lemma is equivalent to the following:

Lemma 2.5 ([31, Lemma 2], Hoppe [36, Theorem 4]) $A_1 - 1$ *has the mixed binomial distribution* $B_N(n - 1, W_1)$. *For a positive integer* $m(< n)$ *and positive integers* $a_1, \ldots, a_m$ $(a_1 + \cdots + a_m < n)$, *under the condition that* $A_1 = a_1, \ldots, A_{m-1} = a_{m-1}$ *are given,* A_m *has the mixed binomial distribution* $B_N(n - a_1 - \cdots - a_{m-1}, W_m)$.

Thus, we have the asymptotic distribution of $(A_1, A_2, \ldots)$.

Proposition 2.6 ([31, 37]) $(A_1, A_2, \ldots)/n$ *converges in distribution to* GEM(θ). *That is,*

$$n^{-1}(A_1, A_2, \ldots) \xrightarrow{d} (p_1, p_2, \ldots) \quad \text{as } n \to \infty.$$

Arratia et al. [37, p. 106–107] proved this result by taking limiting operation, without asking the marginal distribution given by Lemmas 2.4 and 2.5.

2.5.3   Donnelly–Tavaré–Griffiths Formula II

Before treating the Donnelly–Tavaré–Griffiths formula II, we prepare the Waring and the Yule distributions, and their bounded distributions. The p.f. of the Waring distribution Wa(τ, α) is

$$P(X = x) = (\tau - \alpha)\frac{\alpha^{[x]}}{\tau^{[x+1]}}, \quad x = 0, 1, 2, \ldots,$$

where τ, α are positive constants such that $\tau > \alpha$. The p.f. of the Yule distribution Yu(ρ) is

$$P(Y = y) = \rho\frac{(y - 1)!}{(1 + \rho)^{[y]}}, \quad y = 1, 2, \ldots,$$

where $\rho > 0$ (Johnson et al. [38]).

Summing the probabilities $P(X = x)$ over $x = n, n + 1, \ldots$, we obtain the bounded Waring distribution BWa($n; \tau - \alpha$) given by

$$P(X = x) = \begin{cases} (\tau - \alpha)\alpha^{[x]}/\tau^{[x+1]}, & x = 0, 1, 2, \ldots, n - 1, \\ \alpha^{[n]}/\tau^{[n]}, & x = n. \end{cases} \tag{2.19}$$

Summing the probabilities $P(Y = y)$ over $y = n, n + 1, \ldots$, we obtain the bounded Yule distribution BYu($n; \rho$) given by

$$P(Y = y) = \begin{cases} \rho(y - 1)!/(1 + \rho)^{[y]}, & y = 1, 2, \ldots, n - 1, \\ (n - 1)!/(1 + \rho)^{[n-1]}, & y = n. \end{cases} \tag{2.20}$$

The marginal distribution of $(D_1, D_2, \ldots, D_k)$ having the Donnelly–Tavaré–Griffiths formula II (2.14) was given by Donnelly and Tavaré [39] and Yamato [22].

Proposition 2.7 ([22, 39]) *Let r be a positive integer such that $1 \le r \le n - 1$. Then, the p.f. of $D_1, D_2, \ldots, D_r$ is given by*

$$P(D_1 = d_1, D_2 = d_2, \ldots, D_r = d_r)$$

$$= \frac{\theta^r}{(\theta + 1)^{[d(r)]}} \cdot \frac{d(r)!}{d_1(d_1 + d_2) \cdots (d_1 + \cdots + d_r)} \quad (2.21)$$

for $d_1, d_2, \ldots, d_r (= 1, 2, \ldots, n - 1)$ satisfying $d(r) = d_1 + \cdots + d_r < n$. For $d_1, d_2, \ldots, d_r (= 1, 2, \ldots, n - 1)$ satisfying $d_1 + \cdots + d_r = n$, the probability $P(D_1 = d_1, D_2 = d_2, \ldots, D_r = d_r)$ is given by (1.13) with r instead of k.

Thus we have the following:

Corollary 2.2 ([22, 39] and Branson [40]) $n \ge 2$

$$P(D_1 = y) = \begin{cases} \theta(y - 1)!/(\theta + 1)^{[y]} & (y{=}1,2,\ldots,n{-}1) \\ (n - 1)!/(\theta + 1)^{[n-1]} & (y{=}n). \end{cases} \quad (2.22)$$

For $n \ge 2$, D_1 has the bounded Yule distribution $B\mathrm{Yu}(n; \theta)$.

By (2.21), we have the following.

Proposition 2.8 ([22]) *Given $D_1 = d_1, \ldots, D_{r-1} = d_{r-1}$, $D_r - 1$ has the bounded Waring distribution $B\mathrm{Wa}(n - d(r - 1) - 1; \alpha + d(r - 1) + 1, d(r - 1) + 1)$, where $r = 2, \ldots, n - 1, d_1, \ldots, d_{r-1} = 1, 2, \ldots, n - 1$ and $d(r - 1) = d_1 + \cdots + d_{r-1} < n$. Therefore, given $D_1 + \cdots + D_{r-1} = d(r - 1)$, $D_r - 1$ has the bounded Waring distribution $B\mathrm{Wa}(n - d(r - 1) - 1; \alpha + d(r - 1) + 1, d(r - 1) + 1)$, where $r = 2, \ldots, n - 1$ and $r - 1 \le d(r - 1) < n$.*

We put $D(r) = D_1 + \cdots + D_r$ for a positive integer r less than or equal to the number k of distinct partitions in D_n. Then, we have the following.

Proposition 2.9 ([22]) *The p.f of $D(r) = D_{n1} + \cdots + D_{nr}$ is given by*

$$P\big(D(r) = j, r < k\big) = |\, s(j, r)\,| \frac{\alpha^r}{(\alpha + 1)^{[j]}}, \quad j = r, r + 1, \ldots, n - 1,$$

$$P\big(D(r) = n\big) = |\, s(n, r)\,| \frac{\alpha^r}{\alpha^{[n]}}.$$

For the asymptotic distributions as $n \to \infty$, by Propositions 2.7, Corollary 2.2, and Propositions 2.8, 2.9, we have the following.

Proposition 2.10 ([22]) *Let r be a positive integer. Then,*

(i) D_1 has the Yule distribution $\mathrm{Yu}(\alpha)$ asymptotically as $n \to \infty$.

(ii) $(D_1, \ldots, D_r)$ has the asymptotic distribution given by

$$P(D_1 = d_1, \ldots, D_r = d_r)$$

$$= \frac{\alpha^r}{(\alpha + 1)^{[d_1 + \cdots + d_r]}} \cdot \frac{(d_1 + \cdots + d_r)!}{d_1(d_1 + d_2) \cdots (d_1 + \cdots + d_r)}, \quad d_1, \ldots, d_r = 1, 2, \ldots$$

(iii) Given $D_1 = d_1, \ldots, D_{r-1} = d_{r-1}$, $D_r - 1$ has the Waring distribution $\mathrm{Wa}(\alpha + d(r-1) + 1, d(r-1) + 1)$ asymptotically, where $d(r-1) = d_1 + \cdots + d_{r-1}$.

(iv) $D(r) = D_1 + \cdots + D_r$ has the asymptotic distribution given by

$$P(D(r) = j) = \begin{bmatrix} j \\ r \end{bmatrix} \frac{\alpha^r}{(\alpha + 1)^{[j]}}, \quad j = r, r + 1, \ldots$$

2.5.4 Number K_n of Distinct Components (1)

In this subsection, we consider the number K_n of distinct components given by (2.8) with (2.7) (Yamato [41]). By (2.7), the mean and variance of K_n are

$$E(K_n) = \sum_{j=1}^{n} \frac{\theta}{\theta + j - 1}, \quad V(K_n) = \sum_{j=1}^{n} \frac{\theta}{\theta + j - 1} - \sum_{j=1}^{n} \frac{\theta^2}{(\theta + j - 1)^2}. \quad (2.23)$$

Thus, approximately it holds that

$$E(K_n), \quad V(K_n) \approx \theta \log n.$$

Since K_n is the sum of independent random variables, K_n converges in distribution to normal distribution which is given in the following.

Proposition 2.11

$$\frac{K_n - \theta \log n}{\sqrt{\theta \log n}} \xrightarrow{d} N(0, 1). \quad (2.24)$$

As a result that is stronger than Proposition 2.11, Arratia and Tavaré [19] gave the following evaluation by the total variation distance d_{TV} between the law (distribution) $\mathcal{L}(K_n)$ of K_n and the Poisson distribution $Po(E(K_n))$:

$$d_{TV}\left(\mathcal{L}(K_n), Po(E(K_n))\right) \asymp \frac{1}{\log n}. \quad (2.25)$$

The total variation distance between the probability distribution Q_1 and Q_2 over $\{0, 1, 2, \ldots\}$ is denoted as follows:

$$d_{TV}(Q_1, Q_2) = \frac{1}{2} \sum_{j=0}^{\infty} | Q_1(j) - Q_2(j) | .$$

Evaluation (2.25) shows the asymptotic relation between K_n and the Poisson distribution $Po(E(K_n))$. However, since $\xi_1 = 1$ a.s. for K_n given by (2.8), it is preferable to consider the following expression for K_n such that

$$K_n = 1 + L_n \quad \text{a.s. where} \quad L_n := \xi_2 + \cdots + \xi_n . \tag{2.26}$$

We put

$$\mu_n = \sum_{j=2}^{n} \frac{\theta}{\theta + j - 1}, \quad \mu_{2,n} = \sum_{j=2}^{n} \frac{\theta^2}{(\theta + j - 1)^2} . \tag{2.27}$$

By Barbour and Hall [42, Theorems 1 and 2], for the total variation distance between $\mathcal{L}(L_n)$ and $Po(\mu_n)$, we have

$$\frac{1}{32} \min\{1, \mu_n^{-1}\}\mu_{2,n} \le d_{TV}\big(\mathcal{L}(L_n), Po(\mu_n)\big) \le \mu_n^{-1}(1 - e^{-\mu_n})\mu_{2,n} . \tag{2.28}$$

For the total variation distance between $Po(\mu_n)$ and $Po(\lambda)$, by (5), (8), and (9) of Yanaros [43], we have

$$1 - \zeta \le d_{TV}\big(Po(\mu_n), Po(\lambda)\big) \le \sqrt{1 - \zeta^2}, \quad \zeta = \exp\left\{ -\frac{1}{2}(\sqrt{\mu_n} - \sqrt{\lambda})^2 \right\}. \tag{2.29}$$

Here, we assume that

$$\theta \log n \to \lambda \text{ as } n \to \infty \text{ and } \theta \to 0 ,$$

which is equivalent to $\mu_n \to \lambda$ as $n \to \infty$ because of $\mu_n \approx \theta \log n$. Then, $\mu_{2,n} \to 0$ as $\theta \to 0$, because of $\mu_{2,n} < \theta^2 \pi^2/6$. Therefore, by (2.28) and (2.29), we have

$$d_{TV}\big(\mathcal{L}(L_n), Po(\mu_n)\big), d_{TV}\big(Po(\mu_n), Po(\lambda)\big) \to 0 \text{ as } \theta \log n \to \lambda .$$

Since $d_{TV}\big(\mathcal{L}(L_n), Po(\lambda)\big) \le d_{TV}\big(\mathcal{L}(L_n), Po(\mu_n)\big) + d_{TV}\big(Po(\mu_n), Po(\lambda)\big),$

$$d_{TV}\big(\mathcal{L}(L_n), Po(\lambda)\big) \to 0 \text{ as } \theta \log n \to \lambda .$$

Therefore, we have $L_n \overset{d}{\to} Po(\lambda)$. Conversely, if $L_n \overset{d}{\to} Po(\lambda)$, then $E(L_n) = \mu_n \to \lambda$ and $V(L_n) = \mu_n - \mu_{2,n} \to \lambda$ which means $\theta \log n \to \lambda$ as $n \to \infty$ and $\theta \to 0$. Thus, by (2.26), we have the following.

Proposition 2.12 ([41]) *For the number K_n of distinct components of ESF(θ),*

$$K_n \overset{d}{\to} 1 + Po(\lambda) \text{ if and only if } \theta \log n \to \lambda \ (n \to \infty \text{ and } \theta \to 0 \ ; \ 0 < \lambda < \infty). \tag{2.30}$$

The convergence $L_n \overset{d}{\to} Po(\lambda)$ is a special case where the sum of independent Bernoulli random variables converge in distribution to the Poisson distribution. For general cases, e.g., see Wang [44, Theorem 3] and Novak [45, pp. 229–230].

The convergence in (2.30) makes a contrast to the well-known Poisson law of small numbers for the sum of independent and identically distributed Bernoulli random variables. The distribution $1 + Po(\lambda)$ is shifted to the right by one for the Poisson distribution $Po(\lambda)$, and is also known as the size-biased version of $Po(\lambda)$ (e.g., see [37, p. 78, Lemma 4.1]).

2.5.5 Number K_n of Distinct Components (2)

In this subsection, we consider the number K_n of distinct components based on (2.13) with (2.11) and (2.12) [17]. Eliminating the last term from the right-hand sides of (2.11) and (2.12), for $1, 2, \ldots, n - 1$, we let

$$B_1 = \sum_{i=1}^{n-1} \xi_i \xi_{i+1} \,, \quad B_j = \sum_{i=1}^{n-j} \xi_i \bar{\xi}_{i+1} \cdots \bar{\xi}_{i+j-1} \xi_{i+j} \quad (j = 2, 3, \ldots, n-1) \,. \quad (2.31)$$

Lemma 2.6 ([17]) *We can write K_n as follows:*

$$K_n = \sum_{j=1}^{n-1} B_j + R_n \,, \quad (2.32)$$

where

$$R_n = 1 - \prod_{j=1}^{n} \bar{\xi}_j \,. \quad (2.33)$$

Because $\bar{\xi}_1 = 0$ a.s., we have $R_n = 1$ a.s., by (2.33). Therefore, from (2.32), we have the following.

Lemma 2.7 ([17])

$$K_n = \sum_{j=1}^{n-1} B_j + 1 \quad \text{a.s.} \quad (2.34)$$

Here, we put

$$B_j(\infty) = \sum_{i=1}^{\infty} \xi_i \bar{\xi}_{i+1} \cdots \bar{\xi}_{i+j-1} \xi_{i+j} \quad (j = 1, 2, \ldots) \,. \quad (2.35)$$

Lemma 2.8 ([16]) *The random variables $B_j(\infty)$ $(j = 1, 2, \ldots)$ are independent and have the Poisson distribution $Po(\theta/j)$.*

We can also prove the sufficient condition of Proposition 2.12, using the above Lemmas 2.7 and 2.8 [17].

2.6 Related Topics

2.6.1 Patterns of Communication

In order to derive a relation equivalent to (2.5), we use the model by Taga and Isii [12]. Their work was done more than 10 years prior to [6, 7], while did not deal with partition of integer.

Consider a social group π and an information source **S**. The **S** communicates an information **I** to each person in π following the Poisson process with intensity λ. In addition, the information **I** is communicated from a person to another in π following Poisson process with intensity μ. Suppose that these two Poisson processes are independent. We assume that n persons accept the information **I** and let label $1, 2, \ldots, n$ for them, in order of their acceptance. Let k persons accept the information **I** directly from the **S**, and their labels be $j_1(= 1) < j_2 < \cdots < j_k$. Let A_1 be the set which consists of j_1 and persons accepting **I** from j_1, A_2 be the set which consists of j_2 and persons accepting **I** from j_2, and so on. $A_1, \ldots, A_k$ are mutually disjoint and $A_1 \cup \cdots \cup A_k = \{1, 2, \ldots, n\}$. Given $A_1 = a_1, A_2 = a_2, \ldots, A_k = a_k$, let T_i $(i = 1, 2, \ldots, k)$ be the time interval which anyone of $A_i(= a_i)$ communicate **I** to another person. Since the sum of independent Poisson processes is the Poisson process with parameter given by the sum of corresponding parameters, T_i $(i = 1, 2, \ldots, k)$ is the Poisson process with intensity $|a_i|\mu$. Let T_0 be the time interval which **S** communicate **I** to another person. Then, we have

$$
\begin{aligned}
&P\big((n + 1) \text{ from } A_i \mid A_1 = a_1, \ldots, A_k = a_k\big) \\
&= P(T_i < T_0, T_1, \ldots, T_{i-1}, T_{i+1}, \ldots, T_k) \\
&= E[\exp\{-(\lambda + (|a_1| + \cdots + |a_{i-1}| + |a_{i+1}| + \cdots + |a_k|)\mu)T_i\}] \\
&= E[\exp\{-(\lambda + (n - |a_i|)\mu)T_i\}] = \frac{|a_i|\mu}{\lambda + n\mu}.
\end{aligned}
$$

By putting $\theta = \lambda/\mu$, we have

$$
P\big((n + 1) \text{ from } A_i \mid A_1 = a_1, \ldots, A_k = a_k\big) = \frac{|a_i|\theta}{\theta + n}. \tag{2.36}
$$

By the similar method, we have

$$P\big((n+1) \text{ from } \mathbf{S} \mid A_1 = a_1, \ldots, A_k = a_k\big) = P(T_0 < T_1, T_2, \ldots, T_k)$$

$$= \frac{\lambda}{\lambda + n\mu} = \frac{\theta}{\theta + n}. \tag{2.37}$$

The relation given by (2.36) and (2.37) is equivalent to (2.5).

2.6.2 Pitman Sampling Formula (1)

We consider again the Chinese restaurant process, which gives the Pitman sampling formula (Pitman [26]).

Chinese restaurant process (I^*): Consider customers $1, 2, \ldots$ arriving sequentially at an initially empty restaurant that has many tables. The first customer sits at the first table. Let K_n be the number of occupied tables by the first n customers and $A_1^*, A_2^*, \ldots$ denote the set of customers occupying the first table, second table, and so on. Then, given $K_n = k$, $|A_1^*| + \cdots + |A_k^*| = n$. Given the first n customers sat at $K_n = k$ tables and $|A_1^*| = |a_1^*|, \ldots, |A_k^*| = |a_k^*|$, the $(n+1)$th customer select a table as follows ($n = 1, 2, \ldots$):

(1) a new table with probability $\dfrac{\theta + k\alpha}{\theta + n}$, $\hspace{2cm}$ (2.38)

(2) the ith table with probability $\dfrac{|a_i^*| - k\alpha}{\theta + n}$ for $i = 1, \ldots, k$. $\hspace{0.5cm}$ (2.39)

For a disjoint partition $a_1^*, a_2^*, \ldots, a_k^*$ such that $a_1^* \cup a_2^* \cup \cdots \cup a_k^* = \{1, 2, \ldots, n\}$, we have

$$P\big((A_1^*, A_2^*, \ldots, A_k^*) = (a_1^*, a_2^*, \ldots, a_k^*)\big) = \frac{\theta^{[k:\alpha]}}{\theta^{[n]}} \prod_{j=1}^{k} (|a_j^*| - 1)!, \tag{2.40}$$

where $1 \le k \le n$ and $\theta^{[k:\alpha]} = \theta(\theta + \alpha) \cdots (\theta + (k-1)\alpha)$ (Yamato et al. [46, Proposition 2] and [26, (16)]). Let $A_1, A_2, \ldots$ be the number of customers occupying the first table, second table, and so on. The probability of $(A_1, \ldots, A_k)$ is given by

$$P(A_1 = a_1, \ldots, A_k = a_k) = \frac{n! \theta^{[k:\alpha]}}{\theta^{[n]}} \prod_{i=1}^{k} \frac{(1-\alpha)^{[a_i-1]} \theta}{(\sum_{j=i}^{k} a_j)(a_i - 1)!}, \tag{2.41}$$

where $1 \le k \le n, a_j > 0, j = 1, \ldots, k, a_1 + \cdots + a_k = n$, and $\theta^{[k:\alpha]} = \theta(\theta + \alpha) \cdots (\theta + (k-1)\alpha)$ ([46, Theorem 1]). This distribution, in case of $\alpha = 0$, is the Donnelly–Tavaré–Griffiths formula I. We put

C_1 be the number of tables occupied by one customer,

C_2 be the number of tables occupied by two customers, and so on.

Then, $\sum_{i=1}^{n} iC_i = n$ and $\sum_{i=1}^{n} C_i = K_n$. The number of ways giving $(c_1, \ldots, c_n)$ such that $\sum_{i=1}^{n} ic_i = n$ and $\sum_{i=1}^{n} c_i = k$ is $n!/\left(\prod_{j=1}^{n} j!^{c_j} c_j!\right)$. Multiplying this number by (2.40), we have the following probability of $(C_1, \ldots, C_n)$ such that

$$P((C_1, \ldots, C_n) = (c_1, \ldots, c_n)) = n! \frac{\theta^{[k:\alpha]}}{\theta^{[n]}} \prod_{j=1}^{n} \left(\frac{(1-\alpha)^{[j-1]}}{j!}\right)^{c_j} \frac{1}{c_j!}, \quad (2.42)$$

where $\sum_{i=1}^{n} ic_i = n$, $\sum_{i=1}^{n} c_i = k$, and $0 \leq \alpha < 1, \theta > -\alpha$. This distribution is well known as the Pitman sampling formula which generalized the Ewens sampling formula ([26, (17)]).

The Chinese Restaurant Process (II*): Let $D_1, D_2, \ldots$ be waiting times of the next occupied table. The probability of $(D_1, \ldots, D_k)$ is given by

$$P(D_1 = d_1, \ldots, D_{k-1} = d_{k-1}) = \frac{\theta^{[k:\alpha]}}{\theta^{[n]}} \prod_{i=1}^{k} \left(1 - i\alpha + \sum_{j=1}^{i-1} d_j\right)^{[d_i-1]}, \quad (2.43)$$

where $d_k = n - (d_1 + \cdots + d_{k-1}) > 0$. This distribution (2.7), in case of $\alpha = 0$, is the Donnelly–Tavaré–Griffiths formula II and associated with Ewens sampling formula. But the distribution (2.43), in case of $\alpha > 0$, is not associated with the Pitman sampling formula.

The distribution of $K_n = \sum_{j=1}^{n} C_j$ is given by

$$P(K_n = k) = \frac{\theta^{[k:\alpha]}}{\theta^{[n]}} \mid C(n, k, \alpha) \mid \alpha^{-k}, \quad k = 1, 2, \ldots, n, \quad (2.44)$$

where $\mid C(n, k, \alpha) \mid = (-1)^{n-k} C(n, k, \alpha)$ and $C(n, k, \alpha)$ is the generalized Stirling numbers ([46–48]). In the next subsection, we consider asymptotic distributions of K_n and $(C_1, \ldots, C_n)$. Since the case of $\alpha = 0$ gives the Ewens sampling formula, we consider only the case of $0 < \alpha < 1$.

2.6.3 Pitman Sampling Formula (2)

In order to derive asymptotic distributions of K_n and $(C_1, \ldots, C_n)$, we use their moments. By (2.38) and (2.39), we have

$$P(K_{n+1} = k+1 \mid K_n = k) = \frac{\theta + k\alpha}{\theta + n}, \quad P(K_{n+1} = k \mid K_n = k) = \frac{n - k\alpha}{\theta + n}. \quad (2.45)$$

We consider the descending factorial moment $EK_n^{(r)} = E\big(K_n(K_n - 1) \cdots (K_n - r + 1)\big)$ $(r = 1, 2, \ldots)$. Applying (2.45) to

$$
\begin{aligned}
E[K_{n+1}^{(r)} \mid K_n] &= E\big[\big(K_n + (K_{n+1} - K_n)\big)^{(r)} \mid K_n\big] \\
&= E[K_n^{(r)} + rK_n^{(r-1)}(K_{n+1} - K_n) \mid K_n],
\end{aligned}
$$

we have the following recursive relation for $EK_n^{(r)}$.

Lemma 2.9 ([49, (2.4)]) *For $r = 1, 2, \ldots$,*

$$
EK_{n+1}^{(r)} = \left(1 + \frac{\alpha r}{n+\theta}\right) EK_n^{(r)} + \frac{r[\theta + (r-1)\alpha]}{n+\theta} EK_n^{(r-1)}. \tag{2.46}
$$

This relation can be also obtained from [46, (14)] with $f(k) = k^{(r)}$. Using this relation (2.49), we have the following descending factorial moments, which is derived by induction on n and r.

Proposition 2.13 ([49, (2.5)]) *For $n = 1, 2, \ldots$ and $r = 1, 2, \ldots$, the rth descending factorial moment of K_n is*

$$
EK_n^{(r)} = \left(\frac{\theta}{\alpha}\right)^{[r]} \sum_{j=0}^{r} (-1)^{r-j} \binom{r}{j} \frac{(\theta + j\alpha)^{[n]}}{\theta^{[n]}}. \tag{2.47}
$$

For the usual moment EK_n^r, using the Stirling number of the second kind $S(r, i)$ we can write $EK_n^r = \sum_{i=0}^{r} S(r, i)EK_n^{(i)}$. Using (2.47) to this relation, we have the following.

Corollary 2.3 ([49, (2.6)]) *For $n = 1, 2, \ldots$ and $r = 1, 2, \ldots$*

$$
EK_n^r = \sum_{j=0}^{r} (-1)^{r-j} \left(1 + \frac{\theta}{\alpha}\right)^{[j]} R\left(r, j, \frac{\theta}{\alpha}\right) \frac{(\theta + j\alpha + 1)^{[n-1]}}{(\theta + 1)^{[n-1]}}, \tag{2.48}
$$

where $R(r, j, \lambda)$ is the unique function satisfying

$$
\sum_{j=0}^{r} y^{(j)} R(r, j, \lambda) = (y + \lambda)^r \tag{2.49}
$$

for any y, λ and $r = 1, 2, \ldots$.

This function $R(r, j, \lambda)$ was introduced by Carlitz [50] and the relation (2.49) is [49, (3.4)].

The rth moment of K_n given by (2.48) is written as

$$
E[K_n^r] = \sum_{j=0}^{r} (-1)^{r-j} \left(1 + \frac{\theta}{\alpha}\right)^{[j]} R\left(r, j, \frac{\theta}{\alpha}\right) \frac{\Gamma(\theta + j\alpha + n)\Gamma(\theta + 1)}{\Gamma(\theta + n)\Gamma(\theta + j\alpha + 1)}.
$$

By the property of Γ-function that is $\Gamma(\theta + j\alpha + n)/\Gamma(\theta + n) \approx n^{j\alpha}$ and $R(r, r, \theta/\alpha) = 1$ ([50, (3.12)]), we have the following.

Lemma 2.10 ([47, Theorem 3], [49, Lemma 3.1])

$$\mu_r' = \lim_{n \to \infty} E\left[\left(\frac{K_n}{n^\alpha}\right)^r\right] = \left(1 + \frac{\theta}{\alpha}\right)^{[r]} \frac{\Gamma(\theta + 1)}{\Gamma(\theta + 1 + r\alpha)} = \left(\frac{\theta}{\alpha}\right)^{[r]} \frac{\Gamma(\theta)}{\Gamma(\theta + r\alpha)}.$$

This μ_r' is the rth moment of the distribution whose density is given by

$$\frac{\Gamma(\theta + 1)}{\Gamma(\frac{\theta}{\alpha} + 1)} x^{\frac{\theta}{\alpha}} g_\alpha(x), \tag{2.50}$$

where g_α is the density of the Mittag–Leffler distribution with parameter α. The density g_α is the unique function that satisfies

$$\int_0^\infty x^p g_\alpha(x) dx = \frac{\Gamma(p + 1)}{\Gamma(p\alpha + 1)}, \tag{2.51}$$

for any real $p > -1$ ([47], [48, p. 20]).

Let L be the random variable which have the density given by (2.50). Thus, we have the following asymptotic distribution of K_n ([47, Theorem 3], [48, p. 21, ↓ 8]).

Proposition 2.14 ([47, 48]) K_n/n^α *converges as $n \to \infty$ to the distribution with the density given by (3.1), that is, $K_n/n^\alpha \xrightarrow{d} L$.*

For the Pitman sampling formula (2.42), the sum of the right-hand side over all nonnegative integers $c_1, c_2, \ldots, c_n$ satisfying $\sum_{j=1}^n jc_j = n$ is equal to 1. Using this property we have for nonnegative integers $r_1, \ldots, r_n$,

$$E\left[\prod_{j=1}^n C_j^{(r_j)}\right] = \frac{\theta^{[r:\alpha]}(\theta + r\alpha)^{[n-s]}n^{(s)}}{\theta^{[n]}} \prod_{j=1}^n \left(\frac{(1-\alpha)^{[j-1]}}{j!}\right)^{r_j}$$

$$= \frac{(\theta + \alpha)^{[r-1:\alpha]}\Gamma(\theta + 1)}{\Gamma(\theta + r\alpha)} \prod_{j=1}^n \left(\frac{(1-\alpha)^{[j-1]}}{j!}\right)^{r_j} \frac{\Gamma(\theta + r\alpha + n - s)n^{(s)}}{\Gamma(\theta + n)},$$

where $r = r_1 + \cdots + r_n$, $s = \sum_{j=1}^n jr_j \leq n$.

From the above, for a positive integer i and nonnegative integers $r_1, \ldots, r_i$, we also have

$$\lim_{n \to \infty} E\left[\prod_{j=1}^i \left(\frac{j!C_j}{\alpha(1-\alpha)^{[j-1]}n^\alpha}\right)^{r_j}\right] = \mu_{r_1 + \cdots + r_i}', \tag{2.52}$$

where μ_r' is given by Lemma 2.10. Thus, we have the following.

Proposition 2.15 ([49, (4.3)]) *For any positive integer i,*

$$\frac{1}{n^\alpha}(C_1, \ldots, C_i) \xrightarrow{d} L\left(\alpha, \frac{\alpha(1-\alpha)^{[1]}}{2!}, \ldots, \frac{\alpha(1-\alpha)^{[i-1]}}{i!}\right). \tag{2.53}$$

With respect to the right-hand side, the sequence

$$\left(\alpha, \frac{\alpha(1-\alpha)^{[1]}}{2!}, \ldots, \frac{\alpha(1-\alpha)^{[i-1]}}{i!}, \ldots\right)$$

is a discrete probability distribution which is named Sibuya's distribution by Devroye [51] based on [52].

References

1. Ferguson, T.S.: A Bayesian analysis of some nonparametric problems. Ann. Stat. **1**, 209–230 (1973)
2. Sethuraman, J.: A constructive definition of Dirichlet priors. Stat. Sinica **4**, 639–650 (1994)
3. Engen, S.: A note on the geometric series as a species frequency model. Biometrika **62**, 697–699 (1975)
4. MacCloskey, J.W.: A model for the distribution of individuals bu species in an environment. unpublished Ph.D. thesis, Michigan State University (1965)
5. Ewens, W.J.: Population genetics theory—the past and the future. In: Lessard, S. (ed.) Mathematical and Statistical Developments of Evolutionary Theory, pp. 177–227. Kluwer, Amsterdam (1990)
6. Antoniak, C.E.: Mixtures of Dirichlet processes with applications to Bayesian problems. Ann. Stat. **2**, 1152–1174 (1974)
7. Ewens, W.J.: The sampling theory of selectively neutral alleles. Theor. Population Biol. **3**, 87–112 (1972)
8. Johnson, N.L., Kotz, S., Balakrishnan, N.: Discrete Multivariate Distributions. Wiley, New York (1997)
9. Blackwell, D., MacQueen, J.B.: Ferguson distributions via Pólya urn schemes. Ann. Stat. **1**, 353–355 (1973)
10. Aldous, D.J.: Exchangeability and related topics. In: Ecole d'Etéde Probabilités de Saint-Flour XIII—1983. Lecture Notes in Mathematics. vol. 1117. pp. 1–198. Springer, Berlin (1985)
11. Pitman, J.: Combinatorial stochastic processes. In: Ecole d'Etéde Probabilités de Saint-Flour XXXII—2002. Lecture Notes in Mathematics. vol. 1875. Springer, Berlin (2006)
12. Taga, Y., Isii, K.: On a stochastic model concerning the pattern of communication-Diffusion of news in a social group. Ann. Inst. Stat. Math. **11**, 25–43 (1959)
13. Lo, A.Y.: A characterization of the Dirichlet process. Stat. Probab. Lett. **12**, 185–187 (1991)
14. Tavaré, S.: The birth process with immigration, and the genealogical structure of large populations. J. Math. Biol. **25**, 161–168 (1987)
15. Joyce, P., Tavaré, S.: Cycles, permutations and the structure of the Yule process with immigration. Stochast. Process. Appl. **25**, 309–314 (1987)
16. Arratia, R., Barbour, A.D., Tavaré, S.: Poisson processes approximations for the Ewens sampling formula. Ann. Appl. Probab. **2**, 519–535 (1992)
17. Yamato, H.: Asymptotic and approximate discrete distributions for the length of the Ewens sampling formula. In: Proceeding of Pioneering Workshop on Extreme Value and Distribution Theories in Honor of Professor Masaaki Sibuya (to appear)

18. Donnelly, P., Tavaraé, S.: The ages of alleles and coalescent. Adv. Appl. Probab. **18**, 1–19 (1986)
19. Arratia, R., Tavaré, S.: Limit theorems for combinatorial structures via discrete process approximations. Random Struct. Algorithms **3**, 321–345 (1992)
20. Arratia, R., Barbour, A.D., Tavaré, S.: Exploiting the Feller coupling for the Ewens sampling formula [comment on Crane (2016)]. Stat. Sci. **31**, 27–29 (2016)
21. Ethier, S.N.: The infinitely-many-neutral-alleles diffusion model with ages. Adv. Appl. Probab. **22**, 1–24 (1990)
22. Yamato, H.: On the Donnelly-Tavaré-Griffiths formula associated with the coalescent. Commun. Stat. Theory Methods **26**, 589–599 (1997)
23. Nacu, S.: Increments of random partitions. Comb. Probab. Comput. **15**, 589–595 (2006)
24. Young, J.E.: Binary sequential representations of random partitions. Bernoulli **11**, 847–861 (2005)
25. Kingman, J.F.C.: The representation of partitions structures. J. Lond. Math. Soc. **18**, 374–380 (1978)
26. Pitman, J.: Exchangeable and partially exchangeable random partitions. Probab. Theory Relat. Fields **12**, 145–158 (1995)
27. Halmos, P.R.: Random alms. Ann. Math. Stat. **15**, 182–189 (1944)
28. Patil, G.P., Taillie, C.: Diversity as a concept and its implications for random communities. Bull. Int. Stat. Inst. The 47th Session, 497–515 (1977)
29. Gnedin, A.V.: The Bernoulli sieve. Bernoulli **10**, 79–96 (2004)
30. Piman, J.: Random discrete distribution invariant under size-biased permutation. Adv. Appl. Probab. **28**, 525–539 (1996)
31. Sibuya, M., Yamato, H.: Ordered and unordered random partitions of an integer and the GEM distribution. Stat. Probab. Lett. **25**, 177–183 (1995)
32. Yamato, H.: Limiting Bayes estimates of estimable parameters based on Dirichlet processes. J. Jpn. Stat. Soc. **46**, 155–164 (2016)
33. Kingman, J.F.C.: Random discrete distributions. J. R. Stat. Soc. **15**, 1–15 (1975)
34. Kingman, J.F.C.: Poisson Processes. Oxford University Press, New York (1993)
35. Watterson, G.A.: Models for the logarithmic species abundance distributions. Theor. Population Biol. **6**, 217–250 (1974)
36. Hoppe, F.M.: The sampling theory of neutral alleles and an urn model in population genetics. J. Math. Biol **25**, 125–159 (1987)
37. Arratia, R., Barbour, A.D., Tavaré, S.: Logarithmic combinatorial structures: a probabilistic approach. EMS Monographs in Mathematics, EMS Publishing House, Zürich (2003)
38. Johnson, N.L., Kemp, A.W., Kotz, S.: Univariate Discrete Distributions, 3rd edn. Wiley, New York (2005)
39. Donnelly, P. Tavaraé, S. : Chapter2. Chinese Restaurant Process (unpublished lecture notes) (1990)
40. Branson, D.: An urn model and the coalescent in neutral infinite-alleles genetic processes. In: Basawa, I.V., Taylor, R.L. (eds.) Selected Proceedings of the Sheffield Symposium on Applied Probability. IMS Lecture Notes, pp. 174–192 (1991)
41. Yamato, H.: Poisson approximations for sum of Bernoulli random variables and its application to Ewens sampling formula. J. Jpn. Stat. Soc. **47**, 187–195 (2017)
42. Barbour, A.D., Hall, P.: On the rate of Poisson convergence. Math. Proc. Cambridge Philoso. Soc. **95**, 473–480 (1984)
43. Yannaros, N.: Poisson approximation for random sums of Bernoulli random variables. Stat. Probab. Lett. **11**, 161–165 (1991)
44. Wang, Y.H.: On the number of successes in independent trials. Stat. Sinica **3**, 295–312 (1993)
45. Novak, S.Y.: Poisson approximation. Probab. Surv. **16**, 228–276 (2019)
46. Yamato, H., Sibuya, M., Nomachi, T.: Ordered sample from two-parameter GEM distribution. Stat. Probab. Lett. **55**, 19–27 (2001)
47. Piman, J.: Notes on the two parameter generalization of Ewens' random partition structure. Unpublished manuscript

48. Pitman, J.: Brownian motion, bridge, excursion and meander characterized by sampling at independent uniform times. Electron. J. Probab. **4**, 1–33 (1999)
49. Yamato, H., Sibuya, M.: Moments of some statistics of Pitman sampling formula. Bull. Inf. Cybern. **32**, 1–10 (2000)
50. Carlitz, L.: Weighted Stirling numbers of the first and second kind-I. Fibonacci Q. **18**, 147–162 (1980)
51. Devroye, L.: A triptych of discrete distributions related to the stable law. Stat. Probab. Lett. **18**, 349–351 (1993)
52. Sibuya, M.: Generalized hypergeometric, digamma and trigamma distributions. Ann. Inst. Stat. Math. **31**, 373–390 (1979)

Chapter 3
Nonparametric Estimation of Estimable Parameter

Abstract Nonparametric Bayes estimate of estimable parameter is obtained using the Dirichlet process as a priori. By deleting the effect of the prior, we obtain a limit of Bayes estimate. As estimators of estimable parameter, U-statistics and V-statistics are well known. These three statistics are written as a convex combination of U-statistics. For the convex combination, we give H-decomposition. Using this, we give the Berry–Esseen bound of the convex combination in the case of non-degenerate estimable parameter. In the case of degenerate parameter, we give asymptotic distribution of the convex combination. We also give a jackknife statistic of the convex combination of U-statistics.

Keywords Bayes estimate · Berry–Esseen bound · Convex combination of U-statistics · Degenerate · Dirichlet process · Estimable parameter · H-decomposition · Jackknife · Kernel · Nonparametric estimation · Prior · U-statistic · V-statistic

3.1 Nonparametric Bayes Estimator of Estimable Parameter

An estimable parameter or a regular functional of a probability distribution P on $(\mathbb{R}, \mathcal{B})$ is the parameter η such that

$$\eta = \int_{\mathbb{R}^k} g(x_1, x_2, \ldots, x_k) \prod_{i=1}^{k} dP(x_i), \qquad (3.1)$$

where g is a symmetric function of k variables and called the kernel, and the smallest k is called the degree (e.g., see Koroljuk and Borovskich [1], Lee [2]). η is called U-estimable by Lehman and Casella [3].

We consider the case that P is the Dirichlet process DP(θ, Q). By the expression (2.2) of DP(θ, Q), (3.1) can be written as

© The Author(s), under exclusive license to Springer Nature Singapore Pte Ltd. 2020
H. Yamato, *Statistics Based on Dirichlet Processes and Related Topics*,
JSS Research Series in Statistics, https://doi.org/10.1007/978-981-15-6975-3_3

$$
\eta = \sum_{j=1}^{k} \sum_{r_1+\cdots+r_j=k}^{+} \frac{k!}{r_1!\cdots r_j!\,j!}
$$

$$
\times \sum_{i_1\neq\cdots\neq i_j} g\Big(\underbrace{V_{i_1},\ldots,V_{i_1}}_{r_1},\ldots,\underbrace{V_{i_j},\ldots,V_{i_j}}_{r_j}\Big) \times p_{i_1}^{r_1}\cdots p_{i_j}^{r_j}, \qquad (3.2)
$$

where for the given k, $\sum_{r_1+\cdots+r_j=k}^{+}$ denotes the summation over all positive integers $r_1,\ldots,r_j$ $(j=1,\ldots,k)$ satisfying $r_1+\cdots+r_j=k$. The random variables $V_1, V_2, \ldots$ are independent and identically distributed with Q, and independent of the random variables $p_1, p_2, \ldots$. Using (2.17) to the expectation of the right-hand side of (3.2), we have

$$
E[\eta] = \frac{k!}{\theta^{[k]}} \sum_{j=1}^{k} \sum_{r_1+\cdots+r_j=k}^{+} \frac{\theta^j}{r_1\cdots r_j\,j!} \qquad (3.3)
$$

$$
\times \int_{\mathbb{R}^j} g\Big(\underbrace{x_1,\ldots,x_1}_{r_1},\ldots,\underbrace{x_j,\ldots,x_j}_{r_j}\Big) \prod_{i=1}^{j} dQ(x_i),
$$

where we suppose that the existence of the integral of the right-hand side.
This expectation $E[\eta]$ is the non-sample Bayes estimate of η, under the squared error loss.

Let $X_1, X_2, \ldots, X_n$ be a sample of size n from $P(\sim DP(\theta, Q))$. By Theorem 2.1, given $X_1, X_2, \ldots, X_n$, the posterior distribution of P is $DP(\theta+n, \ \hat{P}_n)$, where

$$
\hat{P}_n(\cdot) = \frac{\theta}{\theta+n} Q(\cdot) + \frac{n}{\theta+n} P_n(\cdot), \quad P_n(\cdot) = \frac{1}{n} \sum_{i=1}^{n} \delta_{X_i}(\cdot). \qquad (3.4)
$$

Under the squared error loss, the Bayes estimate of η is given by the conditional expectation $\hat{\eta} = E[\eta \mid X_1, X_2, \ldots, X_n]$ given $X_1, X_2, \ldots, X_n$. Therefore, by (3.3) and (3.4), $\hat{\eta}$ is given by

$$
\hat{\eta} = \frac{k!}{(\theta+n)^{[k]}} \sum_{j=1}^{k} \sum_{r_1+\cdots+r_j=k}^{+} \frac{(\theta+n)^j}{r_1\cdots r_j\,j!}
$$

$$
\times \int_{\mathbb{R}^j} g\Big(\underbrace{x_1,\ldots,x_1}_{r_1},\ldots,\underbrace{x_j,\ldots,x_j}_{r_j}\Big) \prod_{i=1}^{j} d\hat{P}_n(x_i) \qquad (3.5)
$$

(see Yamato [4, (2.2)]). By letting $\theta \to 0$ with Q fixed, we get the limit of Bayes estimate η^* such that

$$\eta^* = \frac{k!}{n^{[k]}} \sum_{j=1}^{k} \sum_{r_1+\cdots+r_j=d}^{+} \frac{n^j}{r_1 \cdots r_j \, j!} \int_{\mathbb{R}^j} g(\underbrace{x_1, \ldots, x_1}_{r_1}, \ldots, \underbrace{x_j, \ldots, x_j}_{r_j}) \prod_{i=1}^{j} dP_n(x_i) .$$

(3.6)

Using the relation $P_n(\cdot) = \sum_{i=1}^{n} \delta_{X_i}(\cdot)/n$ to the right-hand side, we have the following:

$$\eta^* = \binom{n+k-1}{k}^{-1} \sum_{j=1}^{k} \sum_{r_1+\cdots+r_j=k}^{+} \frac{1}{r_1 \cdots r_j \, j!}$$

$$\times \sum_{i_1=1}^{n} \cdots \sum_{i_j=1}^{n} g(\underbrace{X_{i_1}, \ldots, X_{i_1}}_{r_1}, \ldots, \underbrace{X_{i_j}, \ldots, X_{i_j}}_{r_j}).$$

This η^* can be written as follows (see Proposition 2.3 of [4]):

$$\eta^* = \binom{n+k-1}{k}^{-1} \sum_{l=1}^{k} \sum_{\nu_1+\cdots+\nu_l=k}^{+}$$

$$\times \sum_{1 \le i_1 < \cdots < i_l \le n} g(\underbrace{X_{i_1}, \ldots, X_{i_1}}_{\nu_1}, \underbrace{X_{i_2}, \ldots, X_{i_2}}_{\nu_2}, \ldots, \underbrace{X_{i_l}, \ldots, X_{i_l}}_{\nu_l}). \quad (3.7)$$

Hereafter we use B_n instead of η^* and call it *LB*-statistic. Expressing the right-hand side of (3.7) in the equivalent and brief form, we have the following:

Proposition 3.1 ([4, 5])

$$B_n(:= \eta^*) = \binom{n+k-1}{k}^{-1} \sum_{s_1+\cdots+s_n=k} g(\underbrace{X_1, \ldots, X_1}_{s_1}, \ldots, \underbrace{X_n, \ldots, X_n}_{s_n}), \quad (3.8)$$

where $\sum_{s_1+\cdots+s_n=k}$ is the summation of all nonnegative integers $(s_1, \ldots, s_n)$ satisfying $s_1 + \cdots + s_n = k$.

For examples of $\hat{\eta}$ and B_n of degrees 2 and 3, see [5, 6] and Ferguson [7].

3.2   Convex Combination of *U*-Statistics

3.2.1   *Preliminaries*

In this chapter, hereafter, let $X_1, \ldots, X_n$ be a random sample of size n from a distribution P fixed. *LB*-statistic B_n (3.8) is considered as an estimator of the estimable parameter η (3.1), for this sample. As estimators of the estimable parameter η, U-statistic and V-statistic are well known. The idea of U-statistic was introduced by Halmos [8] and U-statistic was named by Hoeffding [9] as an estimator of estimable parameter. The U-statistic is the uniform minimum variance unbiased estimator of parameter in the class of all absolutely continuous distributions. U-statistic is given by

$$U_n = \binom{n}{k}^{-1} \sum_{(n,k)} g(X_{j_1}, \ldots, X_{j_k}),$$

where $\sum_{(n,k)}$ denotes the summation over all subsets $1 \leq i_1, \ldots, i_k \leq n$ of $\{1, 2, \ldots, n\}$. By integrating the kernel of parameter with the empirical distribution function, von Mises [10] introduced a statistic which is well known as V-statistic. V-statistic is given by

$$V_n = \frac{1}{n^k} \sum_{j_1=1}^{n} \cdots \sum_{j_k=1}^{n} g(X_{j_1}, \ldots, X_{j_k}).$$

3.2.2 Convex Combination of U-Statistics

Toda and Yamato [11] introduced a linear combination Y_n of U-statistics which includes these statistics. We renamed it a convex combination of U-statistics. Let $w(r_1, \ldots, r_j; k)$ be a nonnegative and symmetric function of positive integers $r_1, \ldots, r_j$ such that $j = 1, \ldots, k$ and $r_1 + \cdots + r_j = k$, where k is the degree of the kernel g and fixed. We assume that at least one of $w(r_1, \ldots, r_j; k)$'s is positive. We put

$$d(k, j) = \sum_{r_1 + \cdots + r_j = k}^{+} w(r_1, \ldots, r_j; k)$$

for $j = 1, 2, \ldots, k$, where the summation $\sum_{r_1 + \cdots + r_j = k}^{+}$ is taken over all positive integers $r_1, \ldots, r_j$ satisfying $r_1 + \cdots + r_j = k$ with j and k fixed. For $j = 1, \ldots, k$, let $g_{(j)}(x_1, \ldots, x_j)$ be the kernel given by

$$g_{(j)}(x_1, \ldots, x_j) = \frac{1}{d(k, j)} \sum_{r_1 + \cdots + r_j = k}^{+} w(r_1, \ldots, r_j; k)$$
$$\times g(\underbrace{x_1, \ldots, x_1}_{r_1}, \ldots, \underbrace{x_j, \ldots, x_j}_{r_j}). \quad (3.9)$$

Let $U_n^{(j)}$ be the U-statistic associated with this kernel $g_{(j)}(x_1, \ldots, x_j)$ for $j = 1, \ldots, k$. The kernel $g_{(j)}(x_1, \ldots, x_j)$ is symmetric because of the symmetry of $w(r_1, \ldots, r_j; k)$. If $d(k, j)$ is equal to zero for some j, then the associated

$w(r_1, \ldots, r_j; k)$'s are equal to zero. In this case, we let the corresponding statistic $U_n^{(j)}$ be zero.

As an estimator of θ, we consider a convex combination of U-statistics given by

$$Y_n = \frac{1}{D(n, k)} \sum_{j=1}^{k} d(k, j) \binom{n}{j} U_n^{(j)}, \qquad (3.10)$$

where $D(n, k) = \sum_{j=1}^{k} d(k, j) \binom{n}{j}$. Because w's are nonnegative and at least one of them is positive, $D(n, k)$ is positive.

A linear combination of U-statistics, whose special case is V-statistic, was given by Sen [12, (3.2) of p. 319]. However, the statistic Y_n is distinct from it, because a function w determines kernels and coefficients of U-statistics $U_n^{(j)}$ on the right-hand side of (3.12), for a given estimable parameter associated with the kernel g. The statistic Y_n includes U-statistic, V-statistic, and LB-statistic as special cases. We show four examples of the weight function w and the corresponding statistic Y_n.

Example 3.1 (*U-statistic*) First, we consider a function w given by the following:

$$w(1, 1, \ldots, 1; k) = 1$$

and for positive integers $r_1, \ldots, r_j$ ($j = 1, \ldots, k-1$ and $r_1 + \cdots + r_j = k$)

$$w(r_1, \ldots, r_j; k) = 0.$$

Then, $d(k, k) = 1, d(k, j) = 0$ ($j = 1, \ldots, k-1$) and $D(n, k) = \binom{n}{k}$. Thus, the corresponding statistic Y_n is equal to U-statistic U_n.

Example 3.2 (*V-statistic*) We consider a function w given by

$$w(r_1, \ldots, r_j; k) = \frac{k!}{r_1! \cdots r_j!}$$

for positive integers $r_1, \ldots, r_j$ such that $j = 1, \ldots, k$ and $r_1 + \cdots + r_j = k$. The Stirling number of the second kind $S(k, j)$ satisfies the relation $j! S(k, j) = \sum_{r_1 + \cdots + r_j = k}^{+} k!/(r_1! \cdots r_j!)$ (e.g., see Charalambides [13]). Hence, $d(k, j) = j! S(k, j)$ for $j = 1, \ldots, k$. Thus, we have $D(n, k) = \sum_{j=1}^{k} S(k, j)(n)_j = n^k$, because $S(k, j)$ satisfies the relation, $t^k = \sum_{j=1}^{k} S(k, j) t^{(j)}$ where $t^{(j)} = t(t-1) \ldots (t-j+1)$. Therefore, the kernel $g_{(j)}(x_1, \ldots, x_j)$ is equal to

$$\frac{1}{j! S(k, j)} \sum_{r_1 + \cdots + r_j = k}^{+} \frac{k!}{r_1! \cdots r_j!} g(\underbrace{x_1, \ldots, x_1}_{r_1}, \ldots, \underbrace{x_j, \ldots, x_j}_{r_j}).$$

In terms of the U-statistics $U_n^{(j)}$, $j = 1, \ldots, k$ associated with these kernels, the corresponding statistic Y_n is written as

$$\frac{1}{n^k} \sum_{j=1}^{k} S(k, j)(n)_j U_n^{(j)},$$

which is equal to the V-statistic V_n given by (3.10) (e.g., see [1, p. 40], [2, pp. 183–184]).

Example 3.3 (*LB-statistic*) We consider a function w given by

$$w(r_1, \ldots, r_j; k) = 1$$

for positive integers $r_1, \ldots, r_j$ such that $j = 1, \ldots, k$ and $r_1 + \cdots + r_j = k$. For the equation $r_1 + \cdots + r_j = k$ with j and k fixed, the number of its solutions is $\binom{k-1}{j-1}$. Hence, we have $d(k, j) = \binom{k-1}{j-1}$ for $j = 1, \ldots, k$, and $D(n, k) = \sum_{j=1}^{k} \binom{k-1}{j-1}\binom{n}{j} = \binom{n+k-1}{k}$. Thus, the kernel $g_{(j)}(x_1, \ldots, x_j)$ is equal to

$$\binom{k-1}{j-1}^{-1} \sum_{r_1+\cdots+r_j=k}^{+} g(\underbrace{x_1, \ldots, x_1}_{r_1}, \ldots, \underbrace{x_j, \ldots, x_j}_{r_j}).$$

In terms of the U-statistics $U_n^{(j)}$, $j = 1, \ldots, k$ associated with these kernels, the statistic Y_n given by (2.1) is written as

$$\binom{n+k-1}{k}^{-1} \sum_{j=1}^{k} \binom{k-1}{j-1}\binom{n}{j} U_n^{(j)},$$

which is equal to the LB-statistic B_n given by (3.8). For this expression, see Nomachi and Yamato [14, (2.1)]).

Example 3.4 (*S-statistic* [15]) Last, we consider a function w given by

$$w(r_1, \ldots, r_j; k) = \frac{k!}{r_1 \cdots r_j} \tag{3.11}$$

for positive integers $r_1, \ldots, r_j$ such that $j = 1, \ldots, k$ and $r_1 + \cdots + r_j = k$. The Stirling number of the first kind $s(k, j)$ has the expression $j! \mid s(k, j) \mid = \sum_{r_1+\cdots+r_j=k}^{+} k!/(r_1 \cdots r_j)$ (e.g, see [13]). Hence, we have

$$d(k, j) = j! \mid s(k, j) \mid, \quad j = 1, \ldots, k. \tag{3.12}$$

Thus, we have

$$D(n, k) = \sum_{j=1}^{k} | s(k, j) | n^{(j)}. \tag{3.13}$$

Therefore, the kernel $g_{(j)}(x_1, \ldots, x_j)$ is given by

$$g_{(j)}(x_1, \ldots, x_j) = \frac{1}{j! \, | s(k, j) |} \sum_{r_1 + \cdots + r_j = k}^{+} \frac{k!}{r_1 \cdots r_j}$$
$$\times g(\underbrace{x_1, \ldots, x_1}_{r_1}, \ldots, \underbrace{x_j, \ldots, x_j}_{r_j}).$$

In terms of the *U*-statistic $U_n^{(j)}$ associated with these kernels for $j = 1, \ldots, k$, the statistic S_n is given by

$$S_n = \frac{1}{D(n, k)} \sum_{j=1}^{k} d(k, j) \binom{n}{j} U_n^{(j)} = \frac{1}{D(n, k)} \sum_{j=1}^{k} | s(k, j) | n^{(j)} U_n^{(j)}, \tag{3.14}$$

where $D(n, k)$ is given by (3.15). By the recurrence relation of signless Stirling number of the first kind that is $| s(n + 1, k) | = | s(n, k - 1) | + n | s(n, k) |$, $D(n, k)$ satisfies the recurrence relation as follows:

$$D(n, k + 1) = nD(n - 1, k) + kD(n, k), \quad n = k + 1, k + 2, \ldots, \quad k = 1, 2, \ldots$$

For the degree $k = 3, 4$, we have

$$D(n, 3) = n(n^2 + 1), \quad D(n, 4) = n(n^3 + 4n + 1). \tag{3.15}$$

For the degree $k = 1, 2$, the weight function w is same for the *V*-statistic and *S*-statistic and so these two statistics are identical. In general, we cannot write S_n in an explicit form.

For example, we consider probability weighted moment of distribution function F given by $\beta_{r-1} = \int x[F(x)]^{r-1} dF(x)$ ($r = 1, 2, \ldots$). It is also written by

$$\beta_{r-1} = \frac{1}{r} E[\max(X_1, \ldots, X_r)],$$

(e.g., see [2, p. 9]). Its kernel and the kernels associated with Y_n are

$$g(x_1, \ldots, x_r) = \frac{1}{r} \max(x_1, \ldots, x_r),$$
$$g_{(j)}(x_1, \ldots, x_j) = \frac{1}{r} \max(x_1, \ldots, x_j) \text{ for. } j = 1, \ldots, r - 1, \tag{3.16}$$

respectively. Thus, we have

$$U_n = \binom{n}{r}^{-1} \sum_{(n,r)} \frac{1}{r} \max(X_{j_1}, \dots, X_{j_r}) = \frac{1}{r} \binom{n}{r}^{-1} \sum_{i=r}^{n} \binom{i-1}{r-1} X_{(i)},$$

where $X_{(1)}, \dots, X_{(n)}$ are the order statistics of $X_1, \dots, X_n$. Hence,

$$Y_n = \frac{1}{D(n,k)} \sum_{j=1}^{k} d(k,j) \binom{n}{j} U_n^{(j)} = \frac{1}{D(n,k)} \sum_{j=1}^{k} d(k,j) \frac{1}{r} \sum_{i=j}^{n} \binom{i-1}{j-1} X_{(i)}.$$

$$(3.17)$$

Here, we consider the case of $r = 3$ and put

$$\Sigma_{n,\beta}^{(3)} := \frac{1}{3} \sum_{1 \le i_1 < i_2 < i_3 \le n} \max(X_{i_1}, X_{i_2}, X_{i_3}) = \frac{1}{3} \sum_{i=3}^{n} \binom{i-1}{2} X_{(i)},$$

$$\Sigma_{n,\beta}^{(2)} := \frac{1}{3} \sum_{1 \le i_1 < i_2 \le n} \max(X_{i_1}, X_{i_2}) = \frac{1}{3} \sum_{i=2}^{n} \binom{i-1}{1} X_{(i)}.$$

For U-statistic, $d(3,3) = 1$, $d(3,2) = 0$, $d(3,1) = 0$. For LB-statistic, $d(3,3) = 1$, $d(3,2) = 2$, $d(3,1) = 1$. For V-statistic, $d(3,3) = 6$, $d(3,2) = 6$, $d(3,1) = 1$. For S-statistic, $d(3,3) = 6$, $d(3,2) = 6$, $d(3,1) = 2$. Thus, U-statistic, LB-statistic, V-statistic, and S-statistic are written as follows:

$$U_n = \binom{n}{3}^{-1} \Sigma_{n,\beta}^{(3)},$$

$$B_n = \binom{n+2}{3}^{-1} \left\{ \Sigma_{n,\beta}^{(3)} + 2\Sigma_{n,\beta}^{(2)} + \frac{1}{3} \sum_{i=1}^{n} X_i \right\},$$

$$V_n = \frac{1}{n^3} \left\{ 6\Sigma_{n,\beta}^{(3)} + 6\Sigma_{n,\beta}^{(2)} + \frac{1}{3} \sum_{i=1}^{n} X_i \right\},$$

$$S_n = \frac{1}{n(n^2+1)} \left\{ 6\Sigma_{n,\beta}^{(3)} + 6\Sigma_{n,\beta}^{(2)} + \frac{2}{3} \sum_{i=1}^{n} X_i \right\}.$$

3.3   Properties of Convex Combination of U-Statistics

3.3.1   H-Decomposition

We expand the statistic Y_n using the form which is known as H-decomposition in the context of U-statistics, because it is due to Hoeffding (e.g., see [1, 2]).

For the kernel $g_{(j)}(x_1, \ldots, x_j)$ $(j = 1, \ldots, k)$, we put

$$\eta_j = E g_{(j)}(X_1, \ldots, X_j),$$

and

$$\psi_{(j),c}(x_1, \ldots, x_c) = E\big[g_{(j)}(X_1, \ldots, X_j) \mid X_1 = x_1, \ldots, X_c = x_c\big], \quad c = 1, \ldots, j.$$

For $j = 1, \ldots, k$ and $c = 2, 3, \ldots, k$, we put

$$g_{(j)}^{(1)}(x_1) = \psi_{(j),1}(x_1) - \eta_j,$$

$$g_{(j)}^{(c)}(x_1, \ldots, x_c) = \psi_{(j),c}(x_1, \ldots, x_c) - \sum_{i=1}^{c-1} \sum_{1 \le l_1 < \cdots < l_i \le c} g_{(j)}^{(i)}(x_{l_1}, \ldots, x_{l_i}) - \eta_j.$$

Let $H_{(j),n}^{(c)}$ be the U-statistic associated with the kernel $g_{(j)}^{(c)}$, that is,

$$H_{(j),n}^{(c)} = \binom{n}{c}^{-1} \sum_{1 \le l_1 < \cdots < l_c \le n} g_{(j)}^{(c)}(X_{l_1}, \ldots, X_{l_c}). \tag{3.18}$$

Then, the U-statistic $U_n^{(j)}$ associated with the kernel $g_{(j)}$ can be written as follows:

$$U_n^{(j)} = \theta_j + \sum_{c=1}^{j} \binom{j}{c} H_{(j),n}^{(c)}.$$

We note that if $d(k, k) = w(1, \ldots, 1; k) > 0$ then we have $g_{(k)}(x_1, \ldots, x_k) = g(x_1, \ldots, x_k)$ and so $\eta_k = \eta$. Therefore, we shall delete the sub (k). For example, we shall use g and η instead of $g_{(k)}$ and η_k, respectively. We shall also abbreviate ψ_c and $H_n^{(c)}$ instead of $\psi_{(k),c}$ and $H_{(k),n}^{(c)}$, respectively for $c = 1, 2, \ldots, k$. Then, we can write

$$\sqrt{n}(Y_n - \theta) = v_1 + v_2 + v_3,$$

where

$$v_1 = \frac{d(k, k)}{D(n, k)} \binom{n}{k} \sqrt{n}(U_n - \theta),$$

$$v_2 = \frac{d(k, k-1)}{D(n, k)} \binom{n}{k-1} \sqrt{n}(U_n^{(k-1)} - \theta),$$

$$v_3 = \sum_{j=1}^{k-2} \frac{d(k, j)}{D(n, k)} \binom{n}{j} \sqrt{n}(U_n^{(j)} - \theta).$$

In the following, we assume $d(k, k) > 0$. Then, there exists a constant $\delta_k (\geq 0)$ such that

$$\frac{d(k, k)}{D(n, k)} \binom{n}{k} = 1 - \frac{\delta_k}{n} + O\left(\frac{1}{n^2}\right),$$ (3.19)

and

$$\frac{d(k, k-1)}{D(n, k)} \binom{n}{k-1} = \frac{\delta_k}{n} + O\left(\frac{1}{n^2}\right).$$ (3.20)

Therefore, it follows that

$$\sum_{j=1}^{k-2} \frac{d(k, j)}{D(n, k)} \binom{n}{j} = O\left(\frac{1}{n^2}\right).$$

For the U-statistic U_n, $d(k, k)\binom{n}{k}/D(n, k) = 1$ and $\delta_k = 0$. For the V-statistic V_n and the S-statistic S_n, $\delta_k = k(k-1)/2$. For the LB-statistic B_n, $\delta_k = k(k-1)$. (see [15, 16]).

Thus, we obtain the following expansion, which is useful for obtaining the asymptotic properties of Y_n.

Proposition 3.2 ([16, Proposition 2.1]) *We suppose that $d(k, k) > 0$ and*

$$E\left[\, |\, g(X_{j_1}, \ldots, X_{j_k})\,|^2\,\right] < \infty \quad \text{for} \ \ 1 \leq j_1 \leq \cdots \leq j_k \leq k.$$ (3.21)

Then we have

$$\sqrt{n}(Y_n - \eta) = k\left(1 - \frac{\delta_k}{n}\right)\sqrt{n}H_n^{(1)} + \binom{k}{2}\sqrt{n}H_n^{(2)} + \binom{k}{3}\sqrt{n}H_n^{(3)}$$ (3.22)

$$+ \frac{(k-1)\delta_k}{\sqrt{n}}H_{(k-1),n}^{(1)} + \frac{\delta_k}{\sqrt{n}}(\eta_{k-1} - \eta) + R_n,$$

where $E\,|\,R_n\,|^2 = O(n^{-3})$.

3.3.2 Berry–Esseen Bound

Y_n converges to η a.s. That is, under the condition $E\,|\,g(X_{j_1}, \ldots, X_{j_k})\,|^2 < \infty$ for any $j_1, \ldots, j_k$ $(1 \leq j_1 \leq \cdots \leq j_k \leq k)$,

$$Y_n \to \eta \qquad \text{a.s.}$$

Under the condition $\sigma_1^2 = V[g_{(1)}^{(1)}(X_1)] > 0$, in addition to the above, we have the asymptotic normality of Y_n, that is,

$$\sqrt{n}(Y_n - E(Y_n)) \xrightarrow{d} N(0, k^2\sigma_1^2).$$

The following is the Berry–Esseen bound for Y-statistic.

Theorem 3.1 ([11]) *We assume that $\sigma_1^2 > 0$ and*

$$E \mid g(X_1, \ldots, X_k) - \theta \mid^3 \le \xi,$$
$$E \mid g(X_{j_1}, \ldots, X_{j_k}) - Eg(X_{j_1}, \ldots, X_{j_k}) \mid^2 \le \nu_2$$

for any $j_1, \ldots, j_k$ $(1 \le j_1 \le \cdots \le j_k \le k)$, where ξ and ν_2 are positive constants such that $k\sigma_1^2 \le \nu_2 \le \xi^{2/3}$. Then for $n \ge k$

$$\sup_x \left| P\left[\frac{\sqrt{n}(Y_n - EY_n)}{k\sigma_1} \le x \right] - \Phi(x) \right| \le \frac{C\xi}{k^3\sigma_1^3} n^{-\frac{1}{2}}, \qquad (3.23)$$

where Φ is the distribution function of $N(0, 1)$.

For $n \ge k$

$$\sup_x \left| P\left[\frac{\sqrt{n}(Y_n - EY_n)}{\sqrt{Var(Y_n)}} \le x \right] - \Phi(x) \right| \le \frac{C\xi}{k^3\sigma_1^3} n^{-\frac{1}{2}}.$$

3.3.3 Asymptotic Distribution for Degenerate Kernel

For the kernel $g(x_1, \ldots, x_k)$, we put

$$\psi_j(x_1, \ldots, x_j) = E[g(X_1, \ldots, X_k) \mid X_1 = x_1, \ldots, X_j = x_j], \ j = 1, \ldots, k, \qquad (3.24)$$

and

$$\sigma_j^2 = Var[\psi_j(X_1, \ldots, X_j)], \quad j = 1, \ldots, k.$$

If $\sigma_1^2 > 0$, that is, the U-statistic and/or the kernel g is not degenerate, then the statistic Y_n has asymptotic normality. Conversely, we consider the case of $\sigma_1^2 = 0$. For example, U-statistic corresponding to the kernel $g(x_1, x_2) = x_1x_2$ is

$$U_n = \frac{2}{n(n-1)} \sum_{1 \le i < j \le n} X_i X_j = \frac{1}{n-1}\left[\left(\frac{1}{\sqrt{n}} \sum_{i=1}^n X_i \right)^2 - \frac{1}{n} \sum_{i=1}^n X_i^2 \right].$$

Under the distribution satisfying $E(X) = 0$ and $E(X^2) = 1$,

$$nU_n \xrightarrow{d} Z^2 - 1,$$

where Z has a standard normal variable.

In this subsection, we suppose that

$$\sigma_1^2 = \cdots = \sigma_{d-1}^2 = 0 \quad \text{and} \quad \sigma_d^2 > 0,$$

that is, the U-statistic and/or the kernel g is degenerate of order $d - 1$. Hence, $E\psi_d(X_1, \ldots, X_d) = \eta$ and $\psi_1(X_1) = \cdots = \psi_{d-1}(X_1, \ldots, X_{d-1}) = \eta$ a.s.

The asymptotic distribution of U_n with degenerate kernel was described by, for example, [1, 2]. We may summarize their results as follows.

Let W be the Gaussian random measure associated with the distribution F on the real line $(-\infty, \infty)$ such that $EW(A) = 0$ and $EW(A)W(B) = F(A \cap B)$ for any Borel sets A, B. Denote the stochastic integral of the function $f(x_1, \ldots, x_c)$ by $J_c(f)$, that is,

$$J_c(f) = \int_{-\infty}^{\infty} \cdots \int_{-\infty}^{\infty} f(x_1, \ldots, x_c) W(dx_1) \cdots W(dx_c).$$

Alternatively, $J_c(f)$ can be represented by using an orthonormal basis $e_1, e_2, \ldots$ of $L_2(F)$. For any function f_1 and f_2 such that $\int_{R^c} f_i(x_1, \ldots, x_c)^2 \prod_{j=1}^{c} dF(x_j) < \infty$ ($i = 1, 2$), their inner product is given by $(f_1, f_2) = \int_{R^c} f_1(x_1, \ldots, x_c) f_2(x_1, \ldots, x_c) \prod_{j=1}^{c} dF(x_j)$. $J_c(f)$ is also written as

$$J_c(f) = \sum_{i_1=1}^{\infty} \cdots \sum_{i_c=1}^{\infty} (f, e_{i_1} \cdots e_{i_c}) \prod_{l=1}^{\infty} H_{r_l(\mathbf{i})}(Z_l),$$

where H_r is the rth Hermite polynomial, $\{Z_l\}_{l=1}^{\infty}$ is a sequence of independent standard normal random variables, and $r_l(\mathbf{i})$ is the number of indices among $\mathbf{i} = (i_1, \ldots, i_d)$ equal to l.

For the degenerate kernel g of order $d - 1$, under the conditions $E \mid g(X_1, \ldots, X_k) \mid^2 < \infty$, the asymptotic distribution of U_n is given by

$$n^{d/2}(U_n - \eta) \xrightarrow{d} \binom{k}{d} J_d(\xi_{d,d}), \tag{3.25}$$

where $\xi_{d,d}(x_1, \ldots, x_d) = \psi_d(x_1, \ldots, x_d) - \eta$.

The convergence (3.25) also holds under the following conditions (i) or (ii): (i) $E \mid g^{(c)}(X_1, \ldots, X_d) \mid^{2c/(2c-d)} < \infty$ for $c = d, d + 1, \ldots, k$ ([1, Theorem 4.4.2]). (ii) $E \mid g^{(d)}(X_1, \ldots, X_d) \mid^2 < \infty$ and $t^{2c/(2c-d)} P[\mid g^{(c)} \mid > t] \to 0$ ($t \to \infty$) for $c = d + 1, \ldots, k$ (Borovskich [17, Theorem 4.2.3]). We use the convergence (3.27) to Y-statistic.

Using ψ_j given by (3.24), for $1 \le j \le (d-1)/2$, we put

$$\varphi_{d,d-2j}(x_1,\ldots,x_{d-2j}) = E\big[\psi_d(x_1,\ldots,x_{d-2j},X_{d-2j+1},X_{d-2j+1},\ldots,X_{d-j},X_{d-j})\big] \quad (3.26)$$

and

$$w(1^r,2^s;k) = w(\underbrace{1,1,\ldots,1}_{r},\underbrace{2,2,\ldots,2}_{s};k).$$

Then we have the following theorem.

Theorem 3.2 ([18, 19]) *We suppose that*

$$E\big[g(X_{j_1},X_{j_2},\ldots,X_{j_k})^2\big] < \infty \qquad (3.27)$$

for all $j_1, j_2, \ldots, j_k$ such that $1 \le j_1 \le j_2 \le \cdots \le j_k \le k$. We assume $d(k,k) > 0$. Then in case of $d = 2l + 1$ $(l = 1,2,\ldots)$, we have

$$n^{d/2}(Y_n - \eta) \overset{d}{\to} \sum_{j=0}^{l} \tau_j, \qquad (3.28)$$

where

$$\tau_j = \frac{k!}{(k-d)!}\,\frac{1}{(d-2j)!j!}\cdot\frac{w(1^{k-2j},2^j;k)}{w(1^k;k)}J_{d-2j}(\xi_{d,d-2j}), \quad j=0,1,\ldots,l$$

and

$$\xi_{d,d-2j} = \varphi_{d,d-2j}(x_1,\ldots,x_{d-2j}) - \theta, \quad j=0,1,\ldots,l.$$

In case of $d = 2l$ $(l = 1,2,\ldots)$, we have

$$n^{d/2}(Y_n - \eta) \overset{d}{\to} \sum_{j=0}^{l-1} \tau_j + \alpha_l, \qquad (3.29)$$

where

$$\alpha_l = \frac{k!w(1^{k-d},2^l;k)}{(k-d)!l!w(1^k;k)}\big[E\psi_d(X_1,X_1,\ldots,X_l,X_l) - \theta\big].$$

The value $w(1^{k-2j},2^j;k)/w(1^k;k)$ is equal to $1/2^j$ for the V-statistic V_n and the S-statistic S_n, and 1 for the LB-statistic B_n, respectively. We give their asymptotic distributions [20]. l is a positive integer.

For LB-statistic B_n, in case of $d = 2l + 1$, we have

$$n^{d/2}(B_n - \eta) \overset{d}{\to} \frac{k!}{(k-d)!}\sum_{j=0}^{l}\frac{1}{(d-2j)!j!}J_{d-2j}(\xi_{d,d-2j}),$$

where

$$\xi_{d,d-2j}(x_1, \ldots, x_{d-2j}) = \varphi_{d,d-2j}(x_1, \ldots, x_{d-2j}) - \eta \quad (0 \le j \le \frac{d-1}{2}).$$

In case of $d = 2l$,

$$n^{d/2}(B_n - \eta) \xrightarrow{d} \frac{k!}{(k-d)!} \Bigg\{ \sum_{j=0}^{l-1} \frac{1}{(d-2j)!j!} J_{d-2j}(\xi_{d,d-2j})$$

$$+ \frac{1}{l!}[E\psi_d(X_1, X_1, \ldots, X_l, X_l) - \eta] \Bigg\}.$$

For V-statistic V_n, in case of $d = 2l + 1$, we have

$$n^{d/2}(V_n - \theta) \xrightarrow{d} \frac{k!}{(k-d)!} \sum_{j=0}^{l} \frac{1}{(d-2j)!j!2^j} J_{d-2j}(\xi_{d,d-2j}).$$

In case of $d = 2l$,

$$n^{d/2}(V_n - \theta) \xrightarrow{d} \frac{k!}{(k-d)!} \Bigg\{ \sum_{j=0}^{l-1} \frac{1}{(d-2j)!j!2^j} J_{d-2j}(\xi_{d,d-2j})$$

$$+ \frac{1}{l!2^l}[E\psi_d(X_1, X_1, \ldots, X_l, X_l) - \theta] \Bigg\}.$$

For the S-statistic S_n, in case of $d = 2l + 1$, we have

$$n^{d/2}(S_n - \eta) \xrightarrow{d} \frac{k!}{(k-d)!} \sum_{j=0}^{l} \frac{1}{(d-2j)!j!2^j} J_{d-2j}(\xi_{d,d-2j}).$$

In case of $d = 2l$ $(l = 1, 2, \ldots)$, we have

$$n^{d/2}(S_n - \eta) \xrightarrow{d} \frac{k!}{(k-d)!} \Bigg\{ \sum_{j=0}^{l-1} \frac{1}{(d-2j)!j!2^j} J_{d-2j}(\xi_{d,d-2j})$$

$$+ \frac{1}{l!2^l}[E\psi_d(X_1, X_1, \ldots, X_l, X_l) - \eta] \Bigg\}.$$

3.3.4 Corrections to Previous Works

We derived the rate of convergence in distribution of Y_n on some previous works. These works were based on the following inequality from Shorack [21, p.261]: For any random variables W and Δ, we have

$$\sup_x |P(W + \Delta \leq x) - P(W \leq x)| \leq 4(E|W\Delta| + E|\Delta|).$$

Because we have pointed out that the inequality is wrong, we correct the related parts, as follows. Here, we use the Lipschitz condition. That is, for a random variable W, we suppose that there exists a positive constant $M > 0$ and

$$|P(W \leq x) - P(W \leq x')| \leq M|x - x'| \quad (-\infty < x, x' < \infty).$$

Then, we have

$$\sup_x |P(W + \Delta \leq x) - P(W \leq x)| \leq ME|\Delta|.$$

If the distribution of W has a bounded density, the Lipschitz condition is satisfied. [18, p. 68, (3.6)] holds under the Lipschitz condition and the relation of the second line of p.70 ensures that the Lipschitz condition is satisfied for $U_n^{(k-j)}$ in the right parenthesis of (3.10). For [16, p. 81], we assume the Lipschitz condition for $H_n(i)$ ($i = 1, 2, 3$) and $H_{(k-1),n}^{(1)}$. For [22, p. 128], we assume the Lipschitz condition for its U_n^{**} (16). For [23, p. 83], we assume the Lipschitz condition for its U_{n_1,n_2} (4.5).

3.4 Jackknife Statistic of Convex Combination of U-Statistics

We consider jackknife estimator of the estimable parameter η based on the convex combination of U-statistics given by (3.10). For the Y-statistic Y_n, we consider $Y_{n-1}(-i)$ which is the statistic Y_{n-1} based on the sample of size $n - 1$ formed from the original date by deleting the ith data X_i for $i = 1, \ldots, n$. Similarly, we consider $U_{n-1}^{(j)}(-i)$ which is the statistic $U_{n-1}^{(j)}$ based on the sample of size $n - 1$ formed from the original date by deleting the ith data X_i for $i = 1, \ldots, n$ and $j = 1, \ldots, k$. We put

$$\bar{Y}_n = \frac{1}{n} \sum_{i=1}^{n} Y_{n-1}(-i).$$

This is written as follows:

$$\bar{Y}_n = \frac{1}{D(n-1,k)} \sum_{j=1}^{k} d(k,j) \binom{n-1}{j} \times \frac{1}{n} \sum_{i=1}^{n} U_{n-1}^{(j)}(-i)$$

$$= \frac{1}{D(n-1,k)} \sum_{j=1}^{k} d(k,j) \binom{n-1}{j} U_n^{(j)},$$

where we use the relation $U_n^{(j)} = \sum_{i=1}^{n} U_{n-1}^{(j)}(-i)/n$ (for jackknife, e.g., see Shao and Tu [2, 24]). Because a bias-reduced jackknife estimator of η based on Y_n, Y_n^J, is given by $= nY_n - (n-1)\bar{Y}_n$, we have the following.

Proposition 3.3 ([25]) *The bias-reduced jackknife estimator of η based on Y_n, Y_n^J is given by a linear combination of U-statistics such that*

$$Y_n^J = nY_n - (n-1)\bar{Y}_n = \sum_{j=1}^{k} d^J(n,k;j) U_n^{(j)}, \qquad (3.30)$$

where

$$d^J(n,k;j) = \left\{ \frac{n}{D(n,k)} \binom{n}{j} - \frac{n-1}{D(n-1,k)} \binom{n-1}{j} \right\} d(k,j). \qquad (3.31)$$

Since $D(n,k) = \sum_{j=1}^{k} d(k,j) \binom{n}{j}$, we have

$$\sum_{j=1}^{k} d^J(n,k;j) = 1.$$

The statistic Y_n^J given by (2.1) is a linear combination of the one-sample U-statistics and the sum of their coefficients is equal to 1. But some coefficients may be negative as seen later immediately before Proposition 3.4. Therefore, the statistic Y_n^J is not a convex combination of the U-statistics. We can write $Y_n^J = Y_n - b_{Jack}$, where

$$b_{Jack} = (n-1)(\bar{Y}_n - Y_n)$$

is Quenouille's jackknife bias estimator.

For the second central moment of the distribution $P(F)$, we have $U_n = \sum_{i=1}^{n} (X_i - \bar{X})^2/(n-1)$, $V_n = \sum_{i=1}^{n} (X_i - \bar{X})^2/n$, and $B_n = \sum_{i=1}^{n} (X_i - \bar{X})^2/(n+1)$, where $\bar{X} = \sum_{i=1}^{n} X_i$ (see [6]). It is well known that the jackknife statistic V_n^J of this V-statistic V_n is equal to $U_n = \sum_{i=1}^{n} (X_i - \bar{X})^2/(n-1)$ (e.g., see [24]). The jackknife statistic B_n^J of the above LB-statistic B_n is equal to

$$B_n^J = \frac{n+2}{n(n+1)} \sum_{i=1}^{n} (X_i - \bar{X})^2.$$

For the third central moment of the distribution F, we have $U_n = [n/(n-1)(n-2)] \sum_{i=1}^n (X_i - \bar{X})^3$, $V_n = [1/n] \sum_{i=1}^n (X_i - \bar{X})^3$, $S_n = [n/(n^2+1)] \sum_{i=1}^n (X_i - \bar{X})^3$, and $B_n = [n/(n+1)(n+2)] \sum_{i=1}^n (X_i - \bar{X})^3$ ([15]). The jackknife statistic V_n^J of this V-statistic V_n is equal to

$$V_n^J = \frac{n+1}{(n-1)^2} \sum_{i=1}^n (X_i - \bar{X})^3,$$

which is not equal to the U-statistic U_n. The jackknife statistic S_n^J of the above S-statistic S_n is equal to

$$S_n^J = \frac{(n-1)(n-2)(n^2+n+3)}{n(n^2+1)(n^2-2n+2)} \sum_{i=1}^n (X_i - \bar{X})^3.$$

The jackknife statistic B_n^J of the above LB-statistic B_n is equal to

$$B_n^J = \frac{n+6}{(n+1)(n+2)} \sum_{i=1}^n (X_i - \bar{X})^3.$$

Next, we consider jackknife Y-statistics of degrees 2 and 3, not restricting the central moment. In case of $k = 2$,

$$Y_n^J = d^J(n, 2; 2)U_n + d^J(n, 2; 1)U_n^{(1)}, \tag{3.32}$$

where

$$d^J(n, 2; 2) = \frac{d(2, 2)n(n-1)^2}{D(n, 2)D(n-1, 2)} \left\{ w(2; 2) + \frac{(n-2)}{4} w(1, 1; 2) \right\},$$

$$d^J(n, 2; 1) = \frac{d(2, 1)n(n-1)^2}{D(n, 2)D(n-1, 2)} \left\{ w(2; 2) - \frac{1}{2} w(1, 1; 2) \right\}.$$

In order for Y_n^J to be equal to Y_n, at least it must be $d(2, 1) = 0$ or $w(2; 2) = \frac{1}{2}w(1, 1; 2)$. If $d(2, 1) = 0$, then $w(2, 2) = 0$ which means Y_n is U-statistic and $Y_n = Y_n^J$. If $w(2; 2) = \frac{1}{2}w(1, 1; 2)$, then $d^J(n, 2; 2) = 1$ and $Y_n = Y_n^J$. Because $w(2; 2) = \frac{1}{2}w(1, 1; 2)$ is satisfied by V and S-statistics, their jackknife statistics are equal to U-statistic.

On the contrary, in case of $k \geq 3$, the jackknife statistics of V and S-statistics are not equal to U-statistic. For the V-statistic, its jackknife statistic V_n^J is a linear combination of U-statistics having the coefficients such that

$$d^J(n, k; j) = \frac{(n-1)!}{n^{k-1} j!(n-1-j)!} \left\{ \frac{n}{n-j} - \left(\frac{n}{n-1} \right)^{k-1} \right\} \quad (j = 1, \ldots, k).$$

Since $n/(n-1) > 1$, we have $n/(n-1) - [n/(n-1)]^{k-1} \le n/(n-1) - [n/(n-1)]^2 < 0$ and so $d^J(n, k; 1) < 0$. Therefore, V_n^J is not equal to U-statistic for $k \ge 3$. For S-statistic, using $|s(k, 1)| = (k-1)!$ and $|s(k, 2)| = (k-1)! \sum_{i=1}^k (1/i)$,

$$
d^J(n, k; 1) = \left\{ \frac{n^2}{D(n, k)} - \frac{(n-1)^2}{D(n-1, k)} \right\} d(k, 1)
$$

$$
= \frac{d(k, 1)}{D(n, k)D(n-1, k)} \left\{ |s(k, 1)|n^{(2)} - |s(k, 2)|n^{(2)} \right.
$$

$$
\left. - \sum_{j=3}^k |s(k, j)|n^{(j)}[(j-2)n + 1] \right\}
$$

$$
= \frac{d(k, 1)}{D(n, k)D(n-1, k)} \left\{ (k-1)!n^{(2)} \sum_{i=2}^k \frac{1}{i} \sum_{j=3}^k |s(k, j)|n^{(j)}[(j-2)n + 1] \right\}.
$$

Because $d^J(n, k; 1) < 0$, S_n^J is not equal to S-statistic for $k \ge 3$. Thus, we have the following.

Proposition 3.4 *For the degree 2, the jackknife statistics V_n^J and S_n^J based on V_n and S_n, respectively, are equal to the U-statistic U_n. For the degree more than or equal to 3, the jackknife statistics V_n^J and S_n^J are not equal to the U-statistic U_n.*

References

1. Koroljuk, V.S., Borovskich, YuV: Theory of U-statistics. Kluwer Academic Publishers, Dordrecht (1989)
2. Lee, A.J.: U-statistics. Marcel Dekker, New York (1990)
3. Lehmann, E.L., Casella, G.: Theory of Point Estimation, 2nd edn. Springer, New York (1998)
4. Yamato, H.: Limiting Bayes estimate of estimable parameters based on Dirichlet processes. J. Jpn. Stat. Soc. **46**, 155–164 (2016)
5. Yamato, H.: Relation between limiting Bayes estimates and the U-statistics for estimable parameters. J. Jpn. Stat. Soc. **7**, 57–66 (1977)
6. Yamato, H.: Relation between limiting Bayes estimates and the U-statistics for estimable parameters of dgrees 2 and 3. Commun. Stat.-Theory Methods **6**, 55–66 (1977)
7. Ferguson, T.S.: A Bayesian analysis of some nonparametric problems. Ann. Stat. **1**, 209–230 (1973)
8. Halmos, P.R.: The theory of unbiased estimation. Ann. Math. Stat. **17**, 1734–1743 (1946)
9. Hoeffding, W.: A class of statistics with asymptotically normal distribution. Ann. Math. Stat. **19**, 293–325 (1948)
10. von Mises, R.: On the asymptotic distribution of differentiable statistical functions. Ann. Math. Stat. **18**, 309–348 (1947)
11. Toda, K., Yamato. H.: Berry-Esseen bounds for some statistics including LB-statistic and V-statistic. J. Jpn. Stat. Soc. **31**, 225–237 (2001)
12. Sen, P.K.: Some invariance principles relating to jackknifing and their role in sequential analysis. Ann. Stat. **5**, 316–329 (1977)
13. Charalambides, C.A.: Combinatorial Methods in Discrete Distributions. Wiley, New York (2005)

14. Nomachi, T., Yamato, H.: Asymptotic comparisons of U-statistics, V-statistics and limits of Bayes estimates by deficiencies. J. Jpn. Stat. Soc. **31**, 85–98 (2001)
15. Nomachi, T., Kondo, M., Yamato, H.: Higher order efficiency of linear combinations of U-statistics as estimators of estimable parameters. Scientiae Mathematicae Japonicae online **6**, 95–106 (2002)
16. Yamato, H., Nomachi, T., Toda, K.: Edgeworth expansions some statistics including LB-statistic and V-statistic. J. Jpn. Stat. Soc. **33**, 77–94 (2003)
17. Borovskikh, YuV: U-statistics in Banach Spaces. VSP, Utrecht (1996)
18. Yamato, H., Kondo, M.: Rate of convergence in distribution of a linear combination of U-statistics for a degenerate kernel. Bull. Inf. Cybern. **34**, 61–73 (2002)
19. Yamato, H., Kondo, M., Toda, K.: Asymptotic properties of linear combinations of U-statistics with degenerate kernels. J. Nonparametric Stat. **17**, 187–199 (2005)
20. Yamato, H., Toda, K.: Asymptotic distributions of LB-statistics and V-statistics for degenerate kernel. Bull. Inf. Cybern. **33**, 27–42 (2001)
21. Shorack, G.R.: Probability for Statistician. Springer, New York (2000)
22. Yamato, H., Toda, K., Nomachi, T., Maesono, Y.: An Edgeworth expansion of a convex combination of U-statistics based on studentization. Bull. Inf. Cybern. **36**, 105–130 (2004)
23. Toda, K., Yamato, H.: A convex combination of two-sample U-statistics. J. Jpn. Stat. Soc. **36**, 73–89 (2006)
24. Shao, J., Tu, D.: The Jackknife and Bootstrap. Springer, New York (1995)
25. Yamato, H., Toda, K., Nomachi, T.: Jackknifing a convex combination of one-sample U-statistics. Commun. Stat. Theory Methods **36**, 2223–2234 (2007)

Chapter 4
Statistics Related with Ewens Sampling Formula

Abstract For the Ewens sampling formula, the number of distinct components is written as a sum of one and independent Bernoulli random variables almost surely. Therefore, as approximation to the distribution of the number, we consider shifted Poisson and binomial distributions. We recommend the shifted binomial distribution with the almost same first two moments as the number of components and give its two applications. Next, we consider the order statistics of a sample from the GEM distribution on the positive integers. The gaps among the order statistics are independent and have geometric distributions. Therefore, the frequency distribution of order statistics is the Donnelly–Tavaré–Griffiths formula. The number of the largest values has the Yule distribution asymptotically. According to the Erdős–Turán law, log of the order of random permutation is asymptotically normal. This asymptotic property was generalized for the Ewens sampling formula by Barbour and Tavaré. For the formula with random parameter having a continuous distribution, the asymptotic distribution of the law is not normal and related with the distribution of the random parameter.

Keywords Approximation · Bernoulli random variable · Binomial distribution · Charlier polynomial · Donnelly–Tavaré–Griffiths formula · Erdős-Turán law · Ewens sampling formula · Frequency of discrete order statistics · GEM distribution · Krawtchouk polynomial · Mixture of Dirichlet process · Number of distinct observations · Poisson distribution · Shifted distribution · Yule distribution

4.1 Discrete Approximation to Distribution of Number of Components

4.1.1 Preliminaries

For the Ewens sampling formula given by (2.3), the p.f. of the number K_n of distinct observations is given by (2.4). We consider the approximation to its distribution $\mathcal{L}(K_n)$. K_n is written by the sum of Bernoulli random variables such that (2.8) with the independent random variables $\xi_1, \xi_2, \ldots$ having the probabilities (2.7). Therefore, K_n converges in distribution to normal distribution as stated in Proposition 2.11. Using the digamma function $\psi(t) = \Gamma'(t)/\Gamma(t)$ and the trigamma functions $\psi'(t)$, μ_n and $\mu_{2,n}$ given by (2.27) are written as

$$\mu_n = \theta[\psi(\theta + n) - \psi(\theta + 1)], \tag{4.1}$$

$$\mu_{2,n} = \theta[\psi'(\theta + 1) - \psi'(\theta + n)]. \tag{4.2}$$

Therefore, the mean and variance of K_n are written as

$$E(K_n) = 1 + \theta[\psi(\theta + n) - \psi(\theta + 1)], \tag{4.3}$$

$$V(K_n) = \theta[\psi(\theta + n) - \psi(\theta + 1)] - \theta[\psi'(\theta + 1) - \psi'(\theta + n)]. \tag{4.4}$$

Because the digamma and trigamma functions are included in the programming language R, we can draw the approximate p.f. of K_n with the normal distribution having the mean (4.3) and variance (4.4), by simulation using (2.8). This approximation $\mathcal{L}(K_n)$ is better than the approximation $N(\theta \log n, \theta \log n)$ which is based on (2.24). The normal approximation to $\mathcal{L}(K_n)$ is good for not small θ. Since

$$P(K_n = 1) = \frac{(n-1)!}{(\theta + 1) \cdots (\theta + n - 1)} \uparrow 1 \quad \text{as} \quad \theta \downarrow 0,$$

$P(K_n = 1)$ is close to 1 for θ close to zero. As approximation to $\mathcal{L}(K_n)$, the normal approximation is not suitable for θ close to zero. We give examples of normal approximations for $\theta = 0.25, 8$ and $n = 25$, in Figs. 4.1 and 4.2. Because $\mathcal{L}(K_n)$ is discrete distribution, hereafter we consider discrete approximations to $\mathcal{L}(K_n)$. We begin with approximations to sums of Bernoulli random variables (Yamato [1, 2]). Let $\xi_1, \xi_2, \ldots, \xi_n$ be independent Bernoulli random variables with $P(\xi_j = 1) = p_j$ and $P(\xi_j = 0) = 1 - p_j$ $(n = 1, 2, \ldots, n)$. We put

$$S_n = \sum_{j=1}^{n} \xi_j, \quad \lambda_n = \sum_{j=1}^{n} p_j, \quad \lambda_{k,n} = \sum_{j=1}^{n} p_j^k \ (k = 2, 3), \quad \bar{p}_n = \frac{\lambda_n}{n}. \tag{4.5}$$

First, we consider the Poisson approximation to $\mathcal{L}(S_n)$. While we can use $Po(\lambda_n)$ as the approximation, we shall consider a more accurate approximation using the Charlier polynomial. Let $p(x; \lambda) = e^{-\lambda}\lambda^x/x!$ be the p.f. of the Poisson distribution with parameter λ. The jth Charlier polynomial $C_j(x, \lambda)$ is defined by

$$C_j(x, \lambda) = \frac{d^j}{d\lambda^j} p(x, \lambda) \Big/ p(x, \lambda).$$

Fig. 4.1 $n = 25$, $\theta = 0.25$. Bar graph; simulated p.f. of K_n. Dashed line; Normal Approximation

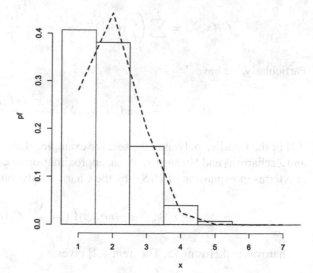

Fig. 4.2 $n = 25$, $\theta = 8$. Bar graph; simulated p.f. of K_n. Dashed line; Normal Approximation

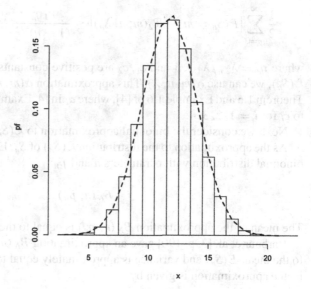

The Charlier polynomials are orthogonal with respect to the p.f. $p(x, \lambda)$. That is,

$$\sum_{x=0}^{\infty} C_k(x, \lambda) C_l(\lambda, x) p(x, \lambda) = \frac{k!}{\lambda^k} \delta_{k,l},$$

where $\delta_{k,l} = 1$ if $k = l$ and $= 0$ if $k \neq l$. The jth Charlier polynomial can be expressed as

$$C_j(x, \lambda) = \sum_{i=0}^{j} \binom{j}{i} (-1)^{j-i} \frac{x(x-1)\cdots(x-i+1)}{\lambda^i}.$$

Particularly, we have

$$C_1(x, \lambda) = \frac{x - \lambda}{\lambda} \quad \text{and} \quad C_2(x, \lambda) = \frac{x^2 - (2\lambda+1)x + \lambda^2}{\lambda^2}.$$

For the Charlier polynomials, see, for example, Barbour et al. [3], Takeuchi [4], and Zacharovas and Hwang [5]. As an approximation to $\mathcal{L}(K_n)$, we consider the first two terms of expansion of $P(S_n)$ by the Charlier polynomial such that

$$\mathcal{P}_2(m; n, \lambda_n) = p(m; \lambda_n)\left(1 - \frac{\lambda_{2,n}}{2} C_2(m, \lambda_n)\right). \tag{4.6}$$

Zacharovas and Hwang [5, Theorem 4.2] gives

$$\frac{1}{2}\sum_{m=0}^{\infty} \left| P(S_n = m) - \mathcal{P}_2(m; n, \lambda_n) \right| \leq \frac{c_1 \eta_n^2}{(1-\eta)^{5/2}} + \frac{c_2 \lambda_{3,n}}{\lambda_n^{3/2}(1-\eta_n)^2}, \tag{4.7}$$

where $\eta_n = \lambda_{2,n}/\lambda_n < 1$ and c_1, c_2 are positive constants. As the approximation to $\mathcal{L}(S_n)$, we can use $\alpha(m, n, \lambda)$. This approximation $\alpha(m, n, \lambda)$ is also obtained from Theorem 1.6 and Example 1.6 of [4], where a_i in [4, Example 1.6] should be changed to c_i for $i = 1, 2, 3, 4$.

Next, we consider the binomial approximation to $\mathcal{L}(S_n)$.

As the approximation to the distribution $\mathcal{L}(S_n)$ of S_n, Ehm [6] gives the following binomial distribution with parameters n and $\bar{p}_n$:

$$B_N(n, \bar{p}_n). \tag{4.8}$$

The mean of the approximation $B_N(n, \bar{p}_n)$ is equal to the mean $E(S_n)$.

Barbour et al. [3, p. 190] gave an approximation $B_N(n', p')$ whose mean is equal to the mean $E(S_n)$ and variance is approximately equal to the variance $V(S_n)$. That is, the approximation is given by

$$B_N(n', p') \quad (n' = \lfloor \lambda_n^2/\lambda_{2,n} \rfloor \text{ and } p' = \lambda_n/n'), \tag{4.9}$$

where $\lfloor x \rfloor$ is an integer close to x. The error of this approximation with the total variation distance is given by [3, Theorem 9.E].

Last, we consider an approximation to $\mathcal{L}(S_n)$ using the Krawtchouk polynomial. Let $p(x; n, p)$ be the p.f. of the binomial distribution $B_N(n, p)$. The Krawtchouk polynomial $L_j^n(x, n, p)$ of degree j is given by

$$L_j^n(x; n, p) = \frac{d^j}{dp^j} p(x; n, p) \Big/ p(x; n, p) = n^{(j)} \Delta^j p(x; n - j, p) \Big/ g(x, n, p),$$

where Δ is the difference operator such that $\Delta^j p(x; n, p) = \Delta^{j-1} p(x - 1; n, p) - \Delta^{j-1} p(x; n, p)$ $(j = 1, 2, \cdots)$ and $\Delta^0 p(x; n, p) = p(x; n, p)$. We can write the factorial moment of S_n as follows:

$$\mu_{(m)} = E[S_n^{(m)}] = \sum_{1 \le j_1 \ne \cdots \ne j_m \le n} p_{j_1} \cdots p_{j_m} \quad (m = 1, 2, \ldots, n),$$

where the summation of the right-hand side is taken over all integers $j_1, \ldots, j_m$ satisfying $1 \le j_1 \ne \cdots \ne j_m \le n$. Let $p_n(k) = \mu_{(k)}/n^{(k)}$, $p = p_n(1) = \bar{p}_n$ and

$$q_n(0) = 1, \quad q_n(j) = \sum_{i=0}^{j} (-1)^i \binom{j}{i} p^i p_n(j - i) \quad (j = 1, 2, \ldots). \tag{4.10}$$

For example, $q_n(1) = 0$, $q_n(2) = p_n(2) - p^2$, $q_n(3) = p_n(3) - 3pp_n(2) + 2p^3$. Then, it holds that

$$P(S_n = x) = p(x; n, p) \left\{ 1 + \sum_{j=2}^{n} \frac{q_n(j)}{j!} L_j^n(x; n, p) \right\},$$

This result was shown by Takeuchi and Takemura [7], using the factorial moment generating function. Using the difference operator, $P(S_n = x)$ is also expressed as

$$P(S_n = x) = p(x; n, p) + \sum_{j=2}^{n} \binom{n}{j} q_n(j) \Delta^j p(x; n - j, p). \tag{4.11}$$

On the other hand, using the generating function, Roos [8] showed the following expression, which is called the Krawtchouk expression by him:

$$P(S_n = x) = \sum_{j=0}^{n} a_j(p) \Delta^j p(x; n - j, p), \tag{4.12}$$

where $a_0(p) = 1$ and

$$a_j(p) = \sum_{1 \le k_1 < \cdots < k_j \le n} \prod_{r=1}^{j} (p_{k_r} - p) \quad (j = 1, \ldots, n).$$

Note that Krawtchouk polynomial $K_j^n(x; n, p)$ defined by [8, (8)] is equal to $p^j(1 - p)^j L_j^n(x; n, p)$. For $s = 0, 1, \ldots, n$, let $\mathcal{B}_s(x; n, p)$ be the finite signed measure such that

$$\mathcal{B}_s(x; n, p) = \sum_{j=0}^{s} a_j(p) \Delta^j p(x; n-j, p) \quad (x = 0, 1, \ldots, n). \tag{4.13}$$

Then for $s = 0, 1, \ldots, n$ and $j = 0, 1, \ldots, s$,

$$\sum_{x=j}^{n} x^{(j)} \mathcal{B}_s(x; n, p) = \mu_{(j)},$$

which means that the first sth moments of $\mathcal{B}_s$ are equal to that of S_n ([8]).

Based on (4.13), we consider approximations to $\mathcal{L}(S_n)$. Since $a_0(p) = 1$ and $a_1(\bar{p}) = 0$, we have $\mathcal{B}_0(x, n, \bar{p}_n) = \mathcal{B}_1(x; n, \bar{p}_n) = p(x; n, \bar{p}_n)$, which is equal to the approximation (4.8). Here we consider $\mathcal{B}_2(x, n, \bar{p}_n)$ which is given by [4, p.84]. Let $\gamma_k(p) = \sum_{j=1}^{n}(p_j - p)^k$ $(k = 1, 2)$. Since $p = \bar{p}_n$, we have $a_1(p) = \gamma_1(p) = 0$ and $a_2(\bar{p}_n) = -\gamma_2(\bar{p}_n)/2$. Then, the approximations are written as follows:

$$\mathcal{B}_2(x; n, \bar{p}_n) = p(x; n, \bar{p}_n) - \frac{\gamma_2(\bar{p}_n)}{2} \Delta^2 p(x; n-2, \bar{p}_n), \tag{4.14}$$

where

$$\Delta^2 p(x; n-2, p) = \frac{p(x; n, p)}{n(n-1)p^2(1-p)^2} \left\{ x^2 - \left[1 + 2(n-1)p\right]x + n(n-1)p^2 \right\}.$$

The error of this approximation with the total variation distance is given by [8, Theorem 2].

4.1.2 Ewens Sampling Formula

For the logarithmic combinatorial structure, including the ESF, the Poisson approximations to $\mathcal{L}(K_n)$ have been derived by Theorem 5.4, Corollary 5.5 and Remark of Arratia et al. [9] and Theorem 8.15 of [10] in detail. We quote their approximations in the case of the ESF, following their notation. Let $p(x, \eta)$ be the value of the p.f. of the Poisson distribution $Po(\eta)$ at the point x. Let $\tau_n = \theta[\psi(n+1) - \psi(\theta+1)]$, $a_n = -\theta^2 \psi'(\theta+1)$, and

$$\nu_n\{s+1\} = p(s, \tau_n)\left(1 + \frac{a_n}{2\tau_n^2}\left\{(s-\tau_n)^2 - \tau_n\right\}\right),$$

where $\psi(t) = \Gamma'(t)/\Gamma(t)$ and $\psi'(t)$ are the digamma and trigamma functions, respectively. Their approximations to $\mathcal{L}(K_n)$ are as follows:

$$\nu_n, \quad Po(1 + \tau_n) \quad \text{and} \quad 1 + Po(\tau_n). \tag{4.15}$$

These approximations (4.15) distribute around

$$1 + \tau_n = 1 + \theta[\psi(n+1) - \psi(\theta+1)] .$$

On the other hand, because

$$\mu_n := E(L_n) = \sum_{i=2}^{n} \frac{\theta}{\theta + i - 1} = \theta[\psi(\theta+n) - \psi(\theta+1)] , \qquad (4.16)$$

by the relation (2.26), we have

$$E(K_n) = 1 + \mu_n = 1 + \theta[\psi(n+\theta) - \psi(\theta+1)] .$$

The function ψ is a monotone increasing function, because $\psi'(x) > 0$ $(x > 0)$. Therefore, μ_n becomes larger than τ_n as θ (> 1) increases. For $0 < \theta < 1$, we have $-\theta/n < \mu_n - \tau_n < 0$, because $\psi(n) = \psi(n+1) - 1/n$. For small θ, τ_n is almost equal to μ_n. As θ increases, τ_n becomes smaller than μ_n. Because the approximations are considered to be based on the Poisson distributions, the centers and dispersions of (4.15) are smaller than those of K_n for large θ. Thus, we consider approximations in which the centers and dispersions are equal to $E(K_n)$ (Yamato [1, 2, 11, 12]).

First, we consider the Poisson approximation using (4.6). Let

$$\mu_{2,n} = \sum_{i=2}^{n} \left(\frac{\theta}{\theta + i - 1} \right)^2 = \theta^2[\psi'(\theta+1) - \psi'(\theta+n)] . \qquad (4.17)$$

By shifting the approximations to $\mathcal{L}(L_n)$, we obtain the shifted Poisson approximation to $\mathcal{L}(K_n)$, as follows:

$$\text{PoA1} : 1 + Po(\mu_n) \quad \text{and} \quad \text{PoA2} : 1 + p(m, \mu_n)\left(1 - \frac{\mu_{2,n}}{2} C_2(m, \mu_n) \right). \qquad (4.18)$$

The expectations and variances of (4.18) are equal to those of K_n, in contrast to (4.15). We compare the approximations PoA1 and PoA2 with the approximate p.f.s of K_n for $n = 25$ and $\theta = 0.25, 8$. In Figs. 4.3 and 4.4, the approximate p.f.s of K_n are drawn as bar graphs and are simulated using R, based on $K_n = \xi_1 + \cdots + \xi_n$. The approximations PoA1 and PoA2 are drawn as dashed and dotted lines, respectively.

The PoA2 is the better approximation. However, a disadvantage of the PoA2 is that that its tails may be negative (see both tails of PoA2 in Fig. 4.4). This is because PoA2 is a signed measure based on the first two terms of an orthogonal expansion by the Charlier polynomials.

We note that

$$E(L_n) = \mu_n \quad \text{and} \quad Var(L_n) = \mu_n \left(1 - \frac{\mu_{2,n}}{\mu_n} \right) .$$

Fig. 4.3 $n = 25$, $\theta = 0.25$.
Bar graph; simulated p.f. of
K_n. Dashed line; PoA1.
Dotted line; PoA2

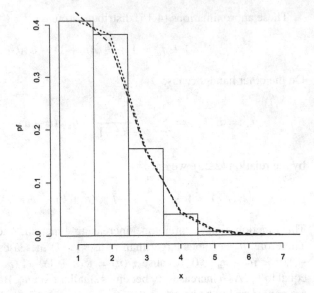

Fig. 4.4 $n = 25$, $\theta = 8$. Bar
graph; simulated p.f. of K_n.
Dashed line; PoA1. Dotted
line; PoA2

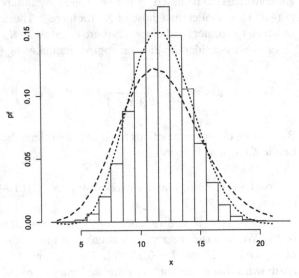

If $\mu_{2,n}/\mu_n$ is small, then $Var(L_n)$ is close to $E(L_n)$ and, therefore, the Poisson distribution is appropriate for approximation to $\mathcal{L}(L_n)$. In general, because $E(L_n) > Var(L_n)$, the binomial distribution may be more appropriate for the approximation to $\mathcal{L}(L_n)$. Therefore, we consider binomial distribution for approximation, obtaining the shifted binomial approximation to $\mathcal{L}(K_n)$.

First, we use (4.9) as an approximation to $\mathcal{L}(L_n)$ and put

$$(n - 1)' = \lfloor \mu_n^2/\mu_{2,n} \rfloor \quad \text{and} \quad p' = \mu_n/(n - 1)' .$$

Then, we have the following shifted binomial approximation to $\mathcal{L}(K_n)$:

$$\text{BnA1} : 1 + B_N((n-1)', p') . \tag{4.19}$$

The corresponding approximate p.f. of K_n is given by

$$g(x-1; (n-1)', p') \quad (x = 1, 2, \ldots, (n-1)' + 1) . \tag{4.20}$$

Next, we use (4.14) as an approximation to $\mathcal{L}(L_n)$ and put

$$\bar{p}_{n-1} = \mu_n/(n-1), \quad \gamma_2(\bar{p}_{n-1}) = \mu_{2,n} - (n-1)\bar{p}_{n-1}^2 .$$

Then, we have the following shifted binomial approximation to $\mathcal{L}(L_n)$:

$$\mathcal{B}_2(n-1, \bar{p}_{n-1})(\{x\}) = p(x; n-1, \bar{p}_{n-1}) - \frac{\gamma_2(\bar{p}_{n-1})}{2} \Delta^2 p(x; n-3, \bar{p}_{n-1}) ,$$

where

$$\Delta^2 p(x; n-3, p)$$
$$= \frac{p(x; n-1, p)}{(n-1)(n-2) p^2 (1-p)^2} \left\{ x^2 - \left[1 + 2(n-2) p\right] x + (n-1)(n-2) p^2 \right\} . \tag{4.21}$$

As an approximation to $\mathcal{L}(K_n)$, we obtain the shifted finite signed measure, as follows:

$$\text{BnA2} : 1 + \mathcal{B}_2(n-1, \bar{p}_{n-1})(\{x\}) , \tag{4.22}$$

which has the same first two moments as K_n. Both (4.20), (4.19) and (4.22) perform well, as shown by the dashed and dotted lines in Figs. 4.5 and 4.6, respectively. However, the tails of (4.22) may be negative, albeit small. This occurs because (4.22) is a signed measure based on the first two terms of an orthogonal expansion, similarly to PoA2 in (4.18). For $n = 25$ and $\theta = 8$, both tails of (4.22) have small negative values, for example, -1.17×10^{-7} and -1.68×10^{-6} for $x = 2$ and 22, respectively. Therefore, the corresponding distribution function is not monotone increasing. Thus, we recommend using the shifted binomial approximation BnA1 (4.20) (4.19) as the approximation to $\mathcal{L}(K_n)$. Two applications of this approximation BnA1 are given in the next section [11, 12].

Fig. 4.5 $n = 25, \theta = 0.25$.
Bar graph; simulated p.f. of
K_n. Dashed line; BnA1.
Dotted line; BnA2

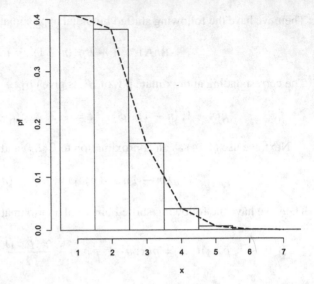

Fig. 4.6 $n = 25, \theta = 8$. Bar
graph; simulated p.f. of K_n.
Dashed line; BnA1. Dotted
line; BnA2

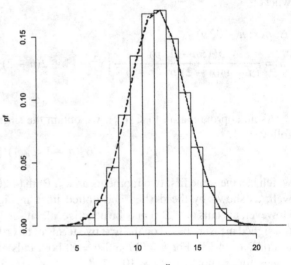

4.1.3 Applications

An approximation to the p.f. of the MLE $\hat{\theta}$ of θ [11, 12]
Given observation $K_n = k$, the MLE $\hat{\theta}$ of the parameter θ is the solution of the
following equation (Ewens [13]):

$$k = \sum_{j=1}^{n} \frac{\theta}{\theta + j - 1} . \tag{4.23}$$

Using the digamma function ψ, (4.23) is written as

$$k = \mu_n(\theta), \quad \mu_n(\theta) = \theta[\psi(\theta + n) - \psi(\theta)] .$$

Because $\mu_n(\theta)$ is a strictly increasing function of θ, for each $k = 1, 2, \ldots, n$, there exists a unique $\mu_n^{-1}(k)$. Thus, we have

$$P(\hat{\theta} = \mu_n^{-1}(k)) = P(K_n = k) \quad (k = 1, 2, \ldots, n) ,$$

or

$$P(\hat{\theta} = x) = P(K_n = \mu_n(x)) \quad (x = \mu_n^{-1}(k), \ k = 1, 2, \ldots, n) .$$

Using (26), the approximation to the p.f. of the MLE $\hat{\theta}$ is given by

$$P(\hat{\theta} = x) \doteq g(k - 1; (n - 1)', p') \quad (\mu_n(x) = k, \ k = 1, 2, \ldots, n) .$$

An estimation of the p.f. of K_n when θ is unknown [11, 12]
The necessary values for the approximation (4.19), (4.20) are

$$\lambda_{n-1} = \sum_{i=2}^{n} \frac{\theta}{\theta + i - 1}, \quad \lambda_{2,n-1} = \sum_{i=2}^{n} \left(\frac{\theta}{\theta + i - 1}\right)^2 = \theta^2[\psi'(\theta + 1) - \psi'(\theta + n)] .$$

As an estimator of θ, we use the MLE $\hat{\theta}$, which satisfies $k = \sum_{j=1}^{n} \theta/(\theta + j - 1)$. Then, from the above relations, we have

$$\lambda_{n-1} = k - 1 , \quad \lambda_{2,n-1}^{**} := \lambda_{2,n-1} = \hat{\theta}^2[\psi'(\hat{\theta} + 1) - \psi'(\hat{\theta} + n)] .$$

We put

$$(n - 1)^{**} = \lfloor (k - 1)^2/\lambda_{2,n-1}^{**} \rfloor, \quad p^{**} = (k - 1)/(n - 1)^{**} .$$

Using the approximation given in (4.20), we obtain the estimator of the p.f. of K_n, as follows:

$$g(x - 1; (n - 1)^{**}, p^{**}) \quad (x = 1, 2, \ldots, (n - 1)^{**} + 1) .$$

4.2 GEM Distribution on Positive Integers

4.2.1 Distribution of Statistics from GEM Distribution

In this subsection, we consider the GEM distribution GEM(θ)) as a random distribution on the positive integers $\mathbf{N} := \{1, 2, \ldots\}$, following Pitman and Yakubovich [14]. Put the values of the distribution of GEM(θ)) as follows:

$$F_0 := 0 \quad \text{and} \quad F_j := \sum_{i=1}^{j} P_i = 1 - \prod_{i=1}^{j}(1 - W_i) \quad (j = 1, 2, \dots).$$

In addition, put

$$N_F(a, b] := \sum_{j=1}^{n} \mathbf{1}(a < F_j \le b) \quad (0 \le a < b < 1),$$

where $\mathbf{1}$ is the indicator function, $\mathbf{1}(A) = 1$ if A occurs, and $= 0$ otherwise. For a sample $U_1, \dots, U_n$ from the uniform distribution $U(0, 1)$, let

$$X_i = N_F(0, U_i) + 1 \quad (i = 1, \dots, n). \tag{4.24}$$

Then, we have the following.

Lemma 4.1 ([14, p. 6])

$$P(X_i = j \mid \mathbf{P_G}) = p_j \quad (i = 1, \dots, n \,; \; j = 1, 2, \dots),$$

That is, we can see $X_1, \dots, X_n$ as a random sample from GEM(θ) on $\mathbf{N}$.

Let $S_0 := 0$ and $S_j := -\log(1 - F_j) = \sum_{i=1}^{j} -\log(1 - W_i)$ $(j = 1, 2, \dots)$, and put

$$N_S(s, t] := \sum_{j=1}^{\infty} \mathbf{1}(s < S_j \le t).$$

Then we have

$$N_S(s, t) = N_F(1 - e^{-s}, 1 - e^{-t}] \quad (0 \ge s < t < \infty). \tag{4.25}$$

We note that $0 < S_1 < S_2 < \dots \uparrow \infty$ a.s., stated in $\downarrow$ 7 of p. 7 of [14]. Since W_i are independent and identically distributed random variables with beta $(1, \theta)$, $-\log(1 - W_i)$ are independent and identically distributed random variables with the exponential distribution $exp(\theta)$. Their sum over $i = 1, \dots, j$ is equal to S_j. Therefore, $N_S(0, t] = \max\{j : S_j \le t\}$ has the following property (e.g., see pp. 297–299 of Billingsley [15]), which is Lemma 2.1 (iii) of [14].

Lemma 4.2 ([14]) N_S *is the Poisson process with intensity θ.*

For the order statistics $X_{1:n} \le \dots \le X_{n:n}$ of the sample $X_1, X_2, \dots, X_n$, we put their gaps as follows:

$$\hat{G}_{n:n} = X_{1:n} - 1, \quad \hat{G}_{i:n} = X_{n+1-i:n} - X_{n-i:n} \quad (i = 1, \dots, n - 1).$$

Let $U_{1:n}, \dots, U_{n:n}$ be the order statistics of the sample $U_1, \dots, U_n$ from $U(0, 1)$. By (3.1) and (3.2),

$$X_{i:n} - 1 = N_F(0, U_{i:n}) = N_F(0, 1 - e^{-\varepsilon_{i:n}}) = N_S(0, \varepsilon_{i:n}),$$

where $\varepsilon_{i:n} = -\log(1 - U_{i:n})$ is the order statistics of the standard exponential distribution $exp(1)$. Since N_S is the Poisson process of intensity θ, we have

$$\hat{G}_{i:n} = N_S(\varepsilon_{n-i:n}, \varepsilon_{n+1-i:n}) \stackrel{d}{=} N_S(0, \varepsilon_{n+1-i:n} - \varepsilon_{n-i:n}) \stackrel{d}{=} Po\left(\theta \frac{\varepsilon}{i}\right), \qquad (4.26)$$

where we use that the difference $\varepsilon_{n+1-i:n} - \varepsilon_{n-i:n}$ has the exponential distribution $exp(1/i)$ and can be equivalently written as ε/i with the standard exponential variable ε. The independence of $\hat{G}_{i:n}$ $(1, \dots, n)$ is obtained from the independence of different intervals of N_S. Then, we have the following.

Proposition 4.1 ([14, Theorem 1.1 and Corollary 1.3]) $\hat{G}_{1:n}, \hat{G}_{2:n}, \dots, \hat{G}_{n:n}$ are *independent and each $\hat{G}_{j:n}$ has the geometric distribution $Ge(j/(j + \theta))$. The maximum $M_n := X_{n:n}$ of the sample can be written as*

$$M_n \stackrel{d}{=} 1 + \hat{G}_{1:n} + \hat{G}_{2:n} + \dots + \hat{G}_{n:n}. \qquad (4.27)$$

Let K_n be the number of distinct values in the sample of size n and $K_{0:n}$ be the total count of values $(1, 2, \dots, M_n)$ that are not observed in the sample. [14, (1.13)] give the following relation:

$$K_n := 1 + \sum_{j=1}^{n-1} \mathbf{1}(\hat{G}_{j:n} > 0). \qquad (4.28)$$

Using the positive part $(x)_+$ of x, we have $x - \mathbf{1}(x > 0) = (x - 1)_+$ and therefore

$$K_{0:n} := M_n - K_n = \hat{G}_{n:n} + \sum_{j=1}^{n-1} (\hat{G}_{j:n} - 1)_+, \qquad (4.29)$$

which is [14, (1.14)], where $(x)_+ = x(x > 0)$, $0(x \le 0)$. The equivalent expression to (3.3) was given by Gnedin et al. [16, (19)]. Since $\hat{G}_{1:n}, \hat{G}_{2:n}, \dots, \hat{G}_{n:n}$ are independent and each $\hat{G}_{i:n}$ has the geometric distribution $Ge(i/(i + \theta))$, we can get the generating function (g.f.) of $K_{0:n}$ as follows:

$$Es^{K_{0:n}} = \frac{n}{n + \theta - \theta s} \prod_{j=1}^{n} \frac{j(j + 2\theta - \theta s)}{(j + \theta)(j + \theta - \theta s)}. \qquad (4.30)$$

By letting $n \to \infty$, we have the g.f. of $K_{0:\infty}$ such that

$$Es^{K_{0:\infty}} = \frac{\Gamma(1 + \theta)\Gamma(1 + \theta - \theta s)}{\Gamma(1 + 2\theta - \theta s)}. \qquad (4.31)$$

Proposition 4.2 ([16, Proposition 5.1]) $K_{0:\infty}$ *has a mixture of Poisson distribution such that*

$$K_{0:\infty} \overset{d}{=} Po(-\theta \log W),$$

where $\overset{d}{=}$ means the equivalence in distribution and W has the beta distribution beta $(1, \theta)$.

It was proved using the g.f. of the Poisson distribution and the Mellin transform of beta $(1, \theta)$. This can be also proved by using the relation among mixed and mixing distributions. It needs an expression of $(\hat{G}_{j:n} - 1)_+$ given by [14, p. 20, ↑ 17–20] as follows:

$$(\hat{G}_{j:n} - 1)_+ \overset{d}{=} N_j\left(\theta B_{q_j} \frac{\varepsilon_j}{j}\right), \tag{4.32}$$

where ε_j are independent standard exponential variables and B_{q_j} are independent Bernoulli variables with parameter q_j, independent also of the ε_j. N_j are independent Poisson variables, independent also of both the ε_j and B_{q_j}.

Thus, by reproductive property of the Poisson distribution, we have

$$(\hat{G}_1 - 1)_+ + \cdots + (\hat{G}_{n-1} - 1)_+ \overset{d}{=} Po\left(\sum_{j=1}^{n-1} \theta B_{q_j} \frac{\varepsilon_j}{j}\right).$$

For the sum of parameters of the right-hand side, it holds that

$$\sum_{j=1}^{\infty} B_{q_j} \frac{\varepsilon_j}{j} \overset{d}{=} -\log W \quad \left(q_j = \frac{\theta}{j+\theta}\right),$$

(see, [14, 7.3]). For the convergence in distribution, the mixed Poisson distribution and its mixing distribution are equivalent (e.g., see Grandell [17, Theorem 2.1(ii)]). Therefore, by the above two relations, we get

$$(\hat{G}_{1:n} - 1)_+ + \cdots + (\hat{G}_{n-1:n} - 1)_+ \overset{d}{\to} Po(-\theta \log W). \tag{4.33}$$

On the other hand, since $E(\hat{G}_{n:n}) = \theta/n$, $V(\hat{G}_{n:n}) = \theta(\theta + n)/n^2 \to 0$ (as $n \to \infty$),

$$\hat{G}_{n:n} \overset{p}{\to} 0, \tag{4.34}$$

where $\overset{p}{\to}$ indicates convergence in probability. Using (4.33) and (4.34) to (4.29), we get Proposition 4.2.

Remark 4.1 (*Bernoulli Sieve*) Gnedin et al. [16, 18] discussed the above problem, under the name *Bernoulli sieve*, by an approach different from [14]. Instead of $\mathbf{P}_G$, GEM(θ) is considered as the following form:

$$Q_0 := 0, \quad Q_j := \prod_{i=1}^{j}(1 - W_i) = 1 - F_j \quad (j = 1, 2, \dots).$$

An allocation of independent random points $U_1, \dots, U_n$ sampled from the uniform distribution $U(0, 1)$ to the intervals consisted of $(p_0(= 0), 1, p_1, p_2, \dots)$ are equivalent to $1 - U_1, \dots, 1 - U_n$ to the intervals consisted of $(Q_0, Q_1, Q_2, \dots)$. Taking logarithm of the latter, it is reduced to an allocation of independent points $e_1, \dots, e_n$ sampled from the standard exponential distribution $exp(1)$ to the intervals consisted of $(0, -\log Q_1, -\log Q_2, \dots)$. The statistic D_j considered by [14, p.1645, ↓] is the number of points $S_j = -\log Q_j$ contained in the intervals of the order statistics of $e_1, \dots, e_n$. That is,

$$D_i = \text{no. of } \{k : S_k \in (\varepsilon_{n-i:n}, \varepsilon_{n-i+1:n})\} \quad (j = 1, 2, \dots, n), \tag{4.35}$$

where $\varepsilon_0 \equiv 0$. In other words, D_i is the number of points of the Poisson process N_S contained in the interval $(\varepsilon_{n-i:n}, \varepsilon_{n-i+1:n})$. Therefore, we can write as follows:

$$D_i = N_S(\varepsilon_{n-i:n}, \varepsilon_{n+1-i:n}).$$

Thus, by (4.26), we have the following.

Proposition 4.3

$$(D_1, D_2, \dots, D_n) \stackrel{d}{=} (\hat{G}_{1:n}, \hat{G}_{2:n}, \dots, \hat{G}_{n:n}).$$

Remark 4.2 Using the probability generating function, [14, Lemma 4.2] derives the distribution of M_n whose g.f. is

$$E(z^{M_n - 1}) = \prod_{i=1}^{n} \frac{i}{i + \theta(1 - z)}. \tag{4.36}$$

We derive the distribution of M_n by the different method. Since

$$M_n - 1 = N_S(0, \varepsilon_{n:n}) = \sum_{i=1}^{\infty} \mathbf{1}(0 < S_i \le \varepsilon_{n:n}),$$

we have

$$P(M_n - 1 \le j) = P(\varepsilon_{n:n} < S_{j+1}) = P(\varepsilon_1, \dots, \varepsilon_n < S_{j+1})$$
$$= \left[\left(1 - e^{-S_{j+1}}\right)^n \right]. \tag{4.37}$$

Since $S_{j+1} = -\sum_{i=1}^{j+1} \log(1 - W_i)$ has the gamma distribution whose density is $\theta(\theta x)^j e^{-\theta x} / j!$, by (4.13), we have

$$P(M_n \le j + 1) = \int_0^\infty (1 - e^{-x})^n \frac{\theta}{j!} (\theta x)^j e^{-\theta x} dx$$

$$= \theta^{j+1} \sum_{r=0}^n \binom{n}{r} (-1)^r \frac{1}{(r + \theta)^{j+1}} .$$

By subtracting $P(M_n \le j)$ from $P(M_n \le j + 1)$, we have the following.

Proposition 4.4 *For* $j = 1, 2, \ldots,$

$$P(M_n = j) = \sum_{r=0}^n \binom{n}{r} (-1)^{r+1} \left(\frac{\theta}{r + \theta}\right)^{j-1} \frac{r}{r + \theta}$$

$$= n\theta^{j-1} \sum_{r=0}^{n-1} \binom{n-1}{r} \frac{(-1)^r}{(r + \theta + 1)^j} . \tag{4.38}$$

To confirm that (4.38) is the p.f. of M_n, we seek the g.f. of $M_n - 1$ based on (4.38).

$$E(z^{M_n-1}) = \sum_{j=1}^\infty z^{j-1} n\theta^{j-1} \sum_{r=0}^{n-1} \binom{n-1}{r} \frac{(-1)^r}{(r + \theta + 1)^j}$$

$$= n \sum_{r=0}^{n-1} (-1)^r \binom{n-1}{r} \frac{1}{r + \theta + 1} \sum_{j=1}^\infty \left(\frac{\theta z}{r + \theta + 1}\right)^{j-1}$$

$$= n \sum_{r=0}^{n-1} \binom{n-1}{r} \frac{(-1)^r}{1 + \theta(1 - z) + r} .$$

Using the relation

$$\sum_{r=0}^n \frac{(-1)^r}{a + r} \binom{n}{r} = \frac{n!}{a^{[n+1]}}$$

(e.g., see [19, ↑ 3]), we have

$$E(z^{M_n-1}) = \frac{n!}{(1 + \theta(1 - z))^{[n]}} ,$$

which is equal to (4.36). Thus, (4.38) is the p.f. of M_n.

4.2.2 Frequency Distribution of Order Statistics from GEM Distribution (1)

For a sample of size n from the GEM distribution on $\mathbf{N}$, assuming that the distinct number of observation is k, let $l_{1:n}$ be the number of the smallest values, $l_{2:n}$ the second smallest values, and so on. Therefore, $l_{k:n}$ is the number of the largest. From Proposition 4.1, $P(\hat{G}_{i:n} = 0) = i/(i + \theta)$ and $P(\hat{G}_{i:n} \neq 0) = \theta/(i + \theta)$, and $\hat{G}_{i:n}$ ($i = 1, \ldots, n$) are independent. This sequence of probability corresponds to (2.6). The number $l_{1:n}$ of smallest values correspond to the the the number A_1 of customers occupying the first table, $l_{2:n}$ correspond to the A_2, and so on. Therefore, the joint probability of $(l_{1:n}, l_{2:n}, \ldots, l_{k:n})$ is given by

$$P\left(l_{1:n} = m_1, l_{2:n} = m_2, \ldots, l_{k:n} = m_k\right)$$
$$= \frac{\theta^k}{\theta^{[n]}} \frac{n!}{(m_1 + m_2 + \cdots + m_k) \cdots (m_{k-1} + m_k)m_k},$$

where $m_1, m_2, \ldots, m_k$ are positive integers satisfying $m_1 + m_2 + \cdots + m_k = n$. Therefore, we have the following.

Proposition 4.5 ([14, (3.4)]) $(l_{1:n}, l_{2:n}, \ldots, l_{k:n})$ has the Donnelly–Tavaré–Griffiths formula I with parameter θ.

Lemmas 2.4, 2.5 and Proposition 2.6 hold for $(l_{1:n}, l_{2:n}, \ldots, l_{k:n})$ instead of $(A_1, A_2, \ldots, A_k)$.

As stated in Proposition 4.1, $\hat{G}_{n:n}$ has the geometric distribution in the case of the GEM distribution. Here, we consider the distribution of $\hat{G}_{n:n}$ in the case that $W_j (j = 1, 2, \ldots)$ are independent but not necessarily identically distributed random variables on $(0, 1)$ for (2.1). We use that $\varepsilon_{1:n}$ is equivalent in distribution to $e(1)/n$, where $e(1)$ is the standard exponential random variables. The p.f. of $\hat{G}_{n:n}$ is given by

$$P(\hat{G}_{n:n} = j) = P(S_j \leq \varepsilon_{1:n} < S_{j+1}) = P(nS_j \leq e(1) < nS_{j+1})$$
$$= E(1 - e^{-S_{j+1}}) - E(1 - e^{-S_j}) \quad (j = 0, 1, \ldots). \quad (4.39)$$

Since $S_j = \sum_{i=1}^{j} -\log(1 - W_i)$ and $W_j (j = 1, 2, \ldots)$ are independent,

$$P(\hat{G}_{n:n} = j) = \prod_{i=1}^{j} E(1 - W_i)^n [1 - E(1 - W_{j+1})^n] \quad (j = 0, 1, \ldots). \quad (4.40)$$

Thus, we have the following.

Proposition 4.6 We consider the distribution of $\hat{G}_{n:n}$ for the residual allocation model (2.1) including the GEM distribution.

(i) *Under the condition that $W_j (j = 1, 2, \ldots)$ are independent but not necessarily identically distributed random variables on $(0, 1)$, the p.f. of $\hat{G}_{n:n}$ is given by (4.40) and its survival function is given by*

$$P(\hat{G}_{n:n} > j) = \prod_{i=1}^{j+1} E[(1 - W_i)^n]. \tag{4.41}$$

(ii) *Under the condition that $W_j (j = 1, 2, \ldots)$ are independent and identically distributed random variables on $(0, 1)$, $\hat{G}_{n:n}$ has the geometric distribution whose p.f. and survival function are given by*

$$P(\hat{G}_{n:n} = j) = (1 - \eta_n)\eta_n^j \quad (j = 0, 1, \ldots), \tag{4.42}$$

$$P(\hat{G}_{n:n} > j) = \eta_n^{j+1} \quad (j = 0, 1, \ldots), \tag{4.43}$$

where $\eta_n = E[(1 - W)^n]$. Especially, if W has beta$(1, \theta)$, then $\eta_n = \theta/(\theta + n)$ and $\hat{G}_{n:n}$ has the geometric distribution $Ge(n/(n + \theta))$.

For the residual allocation model (2.1), we consider the case that $W_i (i = 1, 2, \ldots)$ are independent and each W_i has beta$(1 - \alpha, \theta + i\alpha)$ $(0 < \alpha < 1, \theta > -\alpha)$. This model is called the GEM distribution with parameters (α, θ), abbreviated GEM (α, θ) (e.g., see Pitman [20, Definition 3.3]). The moment of W_i is given by

$$E[(1 - W_i)^n] = \frac{(\theta + i\alpha)^{[n]}}{(\theta + (i - 1)\alpha + 1)^{[n]}}.$$

By Proposition 4.6 (i) , we have the following.

Corollary 4.1 *For GEM(α, β), the p.f. and survival function of $\hat{G}_{n:n}$ are*

$$P(\hat{G}_{n:n} = j) = \frac{\theta^{[j:\alpha]}}{\theta^{[n+1]}} \cdot \frac{(\theta + j\alpha)^{[n+1]}}{(\theta + n + \alpha)^{[j:\alpha]}} \times \left[1 - \frac{(\theta + (j + 1)\alpha)^{[n]}}{(\theta + j\alpha + 1)^{[n]}} \right],$$

$$P(\hat{G}_{n:n} > j) = \frac{\theta^{[j+1:\alpha]}}{\theta^{[n+1]}} \cdot \frac{(\theta + (j + 1)\alpha)^{[n+1]}}{(\theta + n + \alpha)^{[j+1:\alpha]}} \quad (j = 0, 1, \ldots),$$

where $x^{[k:\alpha]} = x(x + \alpha)(x + 2\alpha) \cdots (x + (k - 1)\alpha)$.

4.2.3   Frequency Distribution of Order Statistics from GEM Distribution (2)

For a sample of size n from the GEM distribution on $\mathbf{N}$, assuming that the distinct number of observation is k, let $L_{1:n}$ be the number of the largest values, $L_{2:n}$ the second largest values, and so on. Thus, $L_{k:n}$ is the number of the smallest values. Then, for positive integers $m_1, m_2, \ldots, m_k$ $(m_1 + m_2 + \cdots + m_k = n)$, we have the following relation:

$$
P\big(L_{1:n} = m_1, L_{2:n} = m_2, \ldots, L_{k:n} = m_k\big)
$$
$$
= P\big(l_{1:n} = m_k, l_{2:n} = m_{k-1}, \ldots, l_{k:n} = m_1\big)
$$
$$
= \frac{\theta^k}{\theta^{[n]}} \frac{n!}{m_1(m_1 + m_2)\cdots(m_1 + \cdots + m_k)}, \tag{4.44}
$$

which is equivalent to (2.14). Therefore, we have the following.

Proposition 4.7 $(L_{1:n}.L_{2:n}, \ldots, L_{k:n})$ *has the Donnelly–Tavaré–Griffiths formula II with parameter θ.*

Propositions 2.7, 2.8, 2.9, 2.10, and Corollary 2.2 hold for $(L_{1:n}, L_{2:n}, \ldots, L_{k:n})$ instead of $(D_1, D_2, \ldots, D_k)$.

Especially, L_1 has the bounded Yule distribution $\mathrm{BYu}(n; \theta)$ for $n \geq 2$ and the Yule distribution $\mathrm{Yu}(\alpha)$ asymptotically as $n \to \infty$.

Remark 4.3 Pitman and Yakunovich [14, 21] investigate the asymptotic property of L_1 without referring the Yule distribution. Following [14], here we use L_∞ which is the limiting variable such that $L_1 \overset{d}{\to} L_\infty$ as $n \to \infty$. The right tail probability of L_∞ and its generating function are given by

$$
P(L_\infty > k) = \frac{k!}{(1 + \theta)^{[k]}}, \quad \sum_{k=0}^{\infty} P(L_\infty > k)z^k = {}_2F_1[1, 1; \theta + 1; z],
$$

where ${}_2F_1$ is the Gaussian hypergeometric function. Using the relation of the right-hand side, the expectation and the binomial moments were obtained as follows:

$$
E(L_\infty) = \frac{\theta}{\theta - 1} \text{ for } \theta > 1, \quad \text{and} \quad E\binom{L_\infty}{2} = \frac{\theta}{(\theta - 1)(\theta - 2)} \text{ for } \theta > 2,
$$

$$
E\binom{L_\infty}{3} = \frac{2!\theta}{(\theta - 1)(\theta - 2)(\theta - 3)} \text{ for } \theta > 3.
$$

Pitman and Yakunovich [21] consider the case that $W_j (j = 1, 2, \ldots)$ are independent and identically distributed random variables on $(0, 1)$ for (2.1) such that

$-\log(1 - W)$ is non-lattice with finite mean $\mu_{\log} = E[-\log(1 - W)]$. They give the p.f. of L_∞ as follows:

$$P(L_\infty = m) = \frac{\mu_{m,0}}{m\mu_{\log}},$$

where $\mu_{m,0} = E(W^m)$. In case the GEM distribution GEM(θ), since $\mu_{m,0} = m!/(1 + \theta)^{[m]}$ and $\mu_{\log} = 1/\theta$, L_∞ has the Yule distribution.

4.3 Erdős-Turán Law

We first introduce the Erdős-Turán Law for the Ewens sampling formula following Arratia and Tavaré [22], Barbour and Tavaré [10, 23]. We next show the law for the Ewens sampling formula with random parameter (Yamato [24, 25]).

4.3.1 Erdős–Turán Law for ESF

Let the random partition $C^{(n)} = (C_1, C_2, \ldots, C_n)$ of a positive integer n have the Ewens sampling formula ESF(θ) given by (2.3). In this section, we consider the order $O_n(C^{(n)})$ of the random partition $C^{(n)} = (C_1, \cdots, C_n)$, which is given by

$$O_n(C^{(n)}) = \text{l.c.m.}\{j : 1 \le j \le n, C_j > 0\},$$

where l.c.m. represents the least common multiple. Arratia and Tavaré [22] proved

$$\frac{\log O_n(C^{(n)}) - \frac{\theta}{2} \log^2 n}{\sqrt{\frac{\theta}{3} \log^3 n}} \xrightarrow{d} N(0, 1) \quad \text{as} \quad n \to \infty. \tag{4.45}$$

We show in brief (4.45), following [22, p. 333]: By Lemma 2.8, we have $E(\sum_{j=1}^n B_j(\infty) \log j) \sim (\theta/2) \log^2 n$ and $V(\sum_{j=1}^n B_j(\infty) \log j) \sim (\theta/3) \log^3 n$. Therefore,

$$\frac{\sum_{j=1}^n B_j(\infty) \log j - \frac{\theta}{2} \log^2 n}{\sqrt{\frac{\theta}{3} \log^3 n}} \xrightarrow{d} N[0, 1).$$

For $P_n := \prod_{j=1}^n j^{C_j}$,

$$\frac{|\log P_n - \sum_{j=1}^n B_j(\infty) \log j|}{\sqrt{\frac{\theta}{3} \log^3 n}} \xrightarrow{p} 0. \tag{4.46}$$

Therefore,

$$\frac{\log P_n - \frac{\theta}{2} \log^2 n}{\sqrt{\frac{\theta}{3} \log^3 n}} \xrightarrow{d} N(0, 1). \tag{4.47}$$

Using the relation

$$0 \le \frac{E\left(\log P_n - \log O_n(C^{(n)})\right)}{\log^{3/2} n} = O\left(\frac{(\log \log n)^2}{\log^{1/2} n}\right) \tag{4.48}$$

to (4.47), we have (4.45).

Especially, the case of $\theta = 1$ (random permutation) for (4.45) is well known as the Erdős–Turán law (Erdös and Turán [26]). Gnedin et al. [27] extend the asymptotic normality of $\log O_n$ for sampling from stick-breaking partition of the interval [0, 1], which includes the case of the Ewens sampling formula.

The rate of the convergence (4.45) was given by [10, 23] as follows:

$$\sup_{-\infty < x < \infty} \left| P\left[\left\{ \frac{\theta}{3} \log^3 n \right\}^{-1/2} \left(\log O_n(C^{(n)}) \right. \right. \right.$$
$$\left. \left. \left. - \frac{\theta}{2} \log^2 n + \theta \log n \log \log n \right) \le x \right] - \Phi(x) \right| = O\left(\frac{1}{\log^{1/2} n}\right),$$

where $\Phi(x)$ is the standard normal distribution function.

By using Lemma 2.8 again, we have

$$\frac{\sum_{j=1}^n B_j(\infty) \log j}{\log^2 n} \xrightarrow{p} \frac{\theta}{2}. \tag{4.49}$$

By (4.46), (4.48), and (4.49), we have $\log O_n(C^{(n)}) / \log^2 n \xrightarrow{p} \theta/2$ and therefore

$$\frac{2 \log O_n(C^{(n)})}{\log^2 n} \xrightarrow{p} \theta. \tag{4.50}$$

Hereafter, we consider θ as a positive random variable. Let γ be a distribution of θ. Given θ, let the random discrete distribution P have the Dirichlet process $\mathcal{D}(\theta, Q)$ on $(\mathbb{R}, \mathcal{B})$. Then P has the mixture of Dirichlet processes $\mathcal{D}(\theta, Q)$ with the mixing distribution γ (Antoniak [28]). In other words, we consider the order $O_n(C^{(n)})$ for the random partition $C^{(n)} = (C_1, \cdots, C_n)$ of the integer n based on a sample of size n from the mixture of Dirichlet processes $\mathcal{D}(\theta, Q)$ with the mixing distribution γ of θ.

4.3.2 Erdős–Turán Law for ESF with Random Parameter

Since we consider the case where θ is random variable, the convergence of (4.50) means convergence in distribution, that is,

$$\frac{2 \log O_n(C^{(n)})}{\log^2 n} \xrightarrow{d} \theta.$$

Therefore, we consider the convergence in distribution for $2 \log O_n(C^{(n)})/\log^2 n$ in what follows, instead of $\log O_n(C^{(n)})/\log^2 n$ which converges in distribution to $\theta/2$. For the proofs, see [24, 25]. We assume the distribution function γ of θ has the bonded density g and E_γ denotes the expectation with respect to $\gamma(x)$.

First, we assume $E_\gamma(\theta e^{c_0 \theta}) < \infty$, where c_0 is a positive constant such that $0 < c_0 \le 0.41$. Then, since θ is the positive random variable, we have $E_\gamma \theta^2 < \infty$.

Proposition 4.8 ([24]) *Asymptotically it holds that*

$$\frac{2 \log O_n(C^{(n)})}{\log^2 n} \xrightarrow{d} \theta \quad \text{as} \quad n \to \infty.$$

The rate of convergence is

$$\sup_{-\infty < x, \infty} \left| P\left(\frac{2 \log O_n(C^{(n)})}{\log^2 n} \le x \right) - \gamma(x) \right| = O\left(\frac{1}{\log^{1/3} n} \right). \tag{4.51}$$

The assumptions about θ (γ) are satisfied by the following distributions: (1) The distribution whose support is finite and has the bounded density. (2) The Rayleigh distribution whose density is given by $g(x) = (x/b^2) \exp(-x^2/2b^2)$ ($x > 0$; $b > 0$). (3) The half-normal distribution whose density is given by $g(x) = \sqrt{2}/(\sqrt{\pi}\sigma) \exp[-x^2/(2\sigma^2)]$ ($x > 0$; $\sigma > 0$). (4) The gamma distribution whose density is given by $g(x) = (x/b)^{c-1} e^{-x/b}/b\Gamma(c)$ ($x > 0$; $1/c_0 > b > 0, c > 0$).

Next, we give the evaluation that the rate of convergence is $O(1/\log^{2/5} n)$. Suppose that $E_\gamma(\theta^2)$ exist. For the smoothing lemma (e.g., see [29, Theorem 5.2]) used in the proof of Proposition 4.8, we suppose the following: (i) g be twice differentiable, and $\{xg(x)\}'$ ($= g(x) + xg'(x)$) be of bounded variation; (ii) $g'(x)$ and $xg''(x)$ be bounded, that is, $\{xg(x)\}^{(2)}$ be bounded; and (iii) $g(x) = 0$, $xg'(x) = 0$ for $x = 0$ and $xg'(x) \to 0$ as $x \to +\infty$. Then, we have the following.

Proposition 4.9 ([25])

$$\sup_{-\infty < x < \infty} \left| P\left(\frac{2 \log O_n(C^{(n)})}{\log^2 n} \le x \right) - \left[\gamma(x) + \frac{2}{3 \log n} \{g(x) + xg'(x)\} \right] \right|$$

$$= O\left(\frac{1}{\log^{2/5} n} \right). \tag{4.52}$$

The assumptions about $\theta\,(h)$ are satisfied, for example, by the gamma distribution whose density is $g(x) = x^{c-1}e^{-x}/\Gamma(c)\ (x > 0,\ c > 2)$.

Last, we give the evaluation that the rate of convergence is $O(1/\log^{3/7} n)$. Suppose that $E_\gamma(\theta^3)$ exist. In addition to the previous assumption, we suppose that g is differentiable three times. Suppose that $g'(x)$, $xg^{(2)}(x)$ and $x^2g^{(3)}(x)$ are of bounded variation, and that $g^{(2)}(x)$, $xg^{(3)}(x)$ and $x^2g^{(4)}(x)$ are bounded. Suppose that $xg^{(2)}(x) = 0$, $x^2g^{(3)}(x) = 0$ for $x = 0$ and $xg^{(2)}(x)$, $x^2g^{(3)}(x) \to 0$ as $x \to +\infty$. Then, we have the following.

Proposition 4.10 *([25])*

$$
\sup_x \left| P\left(\frac{2\log O_n(C^{(n)})}{\log^2 n} \le x \right) - \left[\gamma(x) + \frac{2}{3\log n}\{g(x) + xg'(x)\} \right. \right.
$$
$$
\left. \left. + \frac{1}{9\log^2 n}\{6g'(x) + 9xg^{(2)}(x) + 2x^2g^{(3)}(x)\} \right] \right| = O\left(\frac{1}{\log^{3/7} n} \right).
$$

The assumptions about $\theta\,(h)$ of the proposition are satisfied, for example, by the gamma distribution whose density is $g(x) = x^{c-1}e^{-x}/\Gamma(c)\ (x > 0,\ c > 3)$.

References

1. Yamato, H.: Poisson approximations for sum of Bernoulli random variables and its application to Ewens sampling formula. J. Jpn. Stat. Soc. **47**(2), 187–195 (2017)
2. Yamato, H.: Shifted binomial approximation for the Ewens sampling formula. Bull. Inf. Cybern. **49**, 81–88 (2017)
3. Barbour, A.D., Holst, L., Janson, S.: Poisson Approximation. Clarendon Press, Oxford (1992)
4. Takeuchi, K.: Approximation of Probability Distributions (in Japanese). Kyouiku Shuppan, Tokyo (1975)
5. Zacharovas, V., Hwang, H.-K.: A Charlier-Parseval approach to Poisson approximation and its applications. Lithuania Math. J. **50**, 88–119 (2010)
6. Ehm, W.: Binomial approximation to the Poisson binomial distribution. Stat. Probab. Lett. **11**, 7–16 (1991)
7. Takeuchi, K., Takemura, A.: On sum of 0–1 random variables I. univariate case. Ann. Inst. Stat. Math. **39**, 85–102 (1987)
8. Roos, B.: Binomial approximation to the Poisson binomial distribution: The Krawtchouk expansion. Theory Probab. Appl. **45**, 258–272 (2001)
9. Arratia, R., Barbour, A.D., Tavaré, S.: The number of components in logarithmic combinatorial structure. Ann. Appl. Probab. **10**, 331–361 (2000)
10. Arratia, R., Barbour, A.D., Tavaré, S.: Logarithmic combinatorial structures: a probabilistic approach. In: EMS Monographs in Mathematics, EMS Publishing House, Zürich (2003)
11. Yamato, H.: Shifted binomial approximation for the Ewens sampling formula (II). Bull. Inf. Cybern. **50**, 43–50 (2018)
12. Yamato, H.: Asymptotic and approximation discrete distributions for the length of the Ewens sampling formula. In: Proceeding of Pioneering Workshop on Extreme Value and Distribution Theories In Honor of Professor Masaaki Sibuya (to appear)
13. Ewens, W.J.: The sampling theory of selectively neutral alleles. Theor. Population Biol. **3**, 87–112 (1972)

14. Pitman, J., Yakubovich, Y.: Extremes and gaps in sampling from a GEM random discrete distribution. Electron. J. Probab. **22**, 1–26 (2017)
15. Billingsley, P.: Probability and Measure. Wiley, New York (1995)
16. Gnedin, A., Iksanov, A., Negadajlov, P., Rösler, U.: The Bernoulli sieve revisited. Ann. Appl. Probab. **19**, 1634–1655 (2009)
17. J. Grandell.: Mixed Poisson Processes. Chapman and Hall, London (1997)
18. Gnedin, A. Iksanov, A., Marynych, A.: The Bernoulli sieve: overview, arXiv:1005.5705 [math.PR] (Submitted on 31 May 2010)
19. Moriguchi, S., Udagawa, K., Hitotumatsu, S.: Mathematical formula II; Series, Fourier analysis (in Japanese). Iwanami Shoten, Tokyou (1957)
20. Pitman, J.: Combinatorial stochastic processes. In: Ecole d'Etéde Probabilités de Saint-Flour XXXII—2002. Lecture Notes in Mathematics, vol. 1875. Springer, Berlin (2006)
21. Pitman, J., Yakubovich, Y.: Gaps and and interleaving of point processes in sampling from a residual allocation model. Bernoulli **25**, 3623–3651 (2019)
22. Arratia, R., Tavaré, S.: Limit theorems for combinatorial structures via discrete process approximations. Random Struct. Algorithms **3**, 321–345 (1992)
23. Barbour, A.D., Tavaré, S.: A rate for the Erdös-Turán law. Comb. Probab. Comput. **3**, 167–176 (1994)
24. Yamato, H.: The Erdős-Turán law for mixtures of Dirichlet processes. Bull. Inf. Cybern. **45**, 59–66 (2013)
25. Yamato, H.: The Erdős-Turán law for mixtures of Dirichlet processes (II). Bull. Inf. Cybern. **45**, 47–51 (2014)
26. Erdős, P., Turán, P.: On some problems of a statistical group theory III. Acta Mathematica Academiae Scientiarum Hungarica **18**, 309–320 (1967)
27. Gnedin, A., Ikasanov, A., Marynych, A.: A generalization of the Erdős-Turán law for the order of random permutation. Comb. Probab. Comput. **21**, 715–733 (2012)
28. Antoniak, C.E.: Mixtures of Dirichlet processes with applications to Bayesian problems. Ann. Stat. **2**, 1152–1174 (1974)
29. Petrov, V.V.: Limit Theorems of Probability Theory. Clarendon Press, Oxford (2004)

Index

B
Bayes estimate, 30
Bayesian, 1
Bernoulli sieve, 12
Berry–Esseen bound, 39
Binomial approximation, 52

C
Charlier polynomial, 50
Chinese restaurant process, 9, 22
Chinese restaurant process (I), 9
Chinese restaurant process (II), 10, 12
Completely random measure, 2, 4
Conjugacy, 3, 4
Consistency, 12
Convex combination of U-statistics, 32, 43

D
Degenerate, 39
Degree, 29
Dirichlet process, 1, 7, 14, 29
Donnelly–Tavaré–Griffiths formula, 11
Donnelly–Tavaré–Griffiths formula I, 15, 22, 65
Donnelly–Tavaré–Griffiths formula II, 16, 17, 23, 67

E
Erdős-Turán Law, 68
Estimable parameter, 29, 32
Ewens sampling formula, 8, 12, 49, 68
Exchangeable, 11

G
GEM distribution, 8, 13, 14, 65, 67
GEM distribution on the positive integers, 59

H
H-decomposition, 36
Hermite polynomial, 40

J
Jackknife, 44
Jackknife estimator, 43
Jackknife statistic, 44

K
Kernel, 29
Krawtchouk polynomial, 52

L
Laplace transform, 2
LB-statistic, 31, 34, 41
Least common multiple, 68
Limit of Bayes estimate, 30
Linear combination of U-statistics, 32, 44
Lipschitz condition, 43

M
Mittag–Leffler distribution, 25
Mixture of Dirichlet process, 69
Mixture of Poisson distribution, 62
MLE, 58

N
Nonparametric inference, 1
Normalized random measure, 2
Number of distinct components, 18, 20
Number of the largest values, 67
Number of the smallest values, 65

O
Order statistics, 61

P
Partition structure, 12
Pattern of communication, 9
Pitman sampling formula, 22, 23
Poisson approximation, 50, 54
Poisson–Dirichlet distribution, 13, 14
Poisson process, 60, 61
Pólya sequence, 9
Posterior distribution, 8, 30
Probability weighted moment, 35
Process neutral to the right, 3, 4

R
Random parameter, 68
Regular functional, 29
Residual allocation model, 12

S
Shifted binomial approximation, 56
Shifted Poisson approximation, 55
Size-biased permutation, 13
Spacing, 10
S-statistic, 34, 41, 46
Stick-breaking model, 12

T
Total variation distance, 18

U
U-estimable, 29
U-statistic, 32, 33

V
V-statistic, 32, 33, 41, 45

W
Waring distribution, 16–18

Y
Y-statistic, 40
Yule distribution, 16, 17, 67